The Grandfather of American Tea:
My Tea Story

Donnie Barrett

Donnie Barrett

ISBN: 978-1-961485-83-9 Paperback

ISBN: 978-1-961485-84-6 Hardback

Copyright ©2025 Donnie Barrett.

FV-7

All rights reserved. No part of this book may be reproduced in any form without written permission from the publisher, except for brief passages included in a review appearing in a newspaper or magazine.

Printed in the United States of America.

https://fairhopeteaplantation.com/

Published by
Intellect Publishing, LLC
www.IntellectPublishing.com

Contents

	Page Number
Foreward	5
Introduction	9
History of the tea plant	15
How I got started	18
Growing tea	29
Pruning tea	36
Tea seeds	41
Plantation?	46
Learning to make black tea	51
Learning to make green tea	65
Cup of chemicals	68
Plucking	71
Side growth	75
Making black tea	80
Making green	92
Yellow tea	101
Drying tea	104
Blending	107

Loose leaf	110
Brewing	113
Pests	117
Tea rolling machine	124
Tea bagger machine	128
Tea tours	131
Tea professionals	137
Shen Hua	139
My life story	143
Conclusion	157
Acknowledgements	161

Foreward

The Grandfather of American Tea

Frederic was no gentleman. That's Hurricane Frederic, the category 4 storm which hit the northern gulf coast in September 1979, leaving destruction and challenges to lives and communities in its path. But as often happens in troubling times, darkness changes to light, lemons can become lemonade, and to the point of this book - tea plants rescued from a hurricane debris pile became inspiring possibilities for Donnie Barrett, a young Fairhope man.

Forty-five years later, Donnie has written the account of what happened to those rescued plants in *The Grandfather of American Tea*. It is a story of the evolution of an idea to grow quality tea plants on five acres in Fairhope, Alabama. And it is about learning the craft of processing tea, which according to Donnie, really can't be taught - although he did go to China to observe tea farmers to help the learning curve!

Today, there are over 61,000 tea plants of the *Fairhope cultivar camellia sinensis,* flourishing at Donnie and Lottie Barrett's Fairhope Tea Plantation. This is one of the oldest tea farms in the United States, one of the most prolific in producing tea, and one famous for the delicious and aromatic black, green and oolong teas enjoyed by

Fairhopers and visitors to the Plantation. A cup of FTP tea is sublime.

The Grandfather of American Tea is not only about the techniques of growing tea and establishing a successful business, but it is an inherently autobiographical book. Throughout, Donnie weaves personal stories, including about growing up on the Auburn University Gulf Coast Substation experimental research farm. Those years prepared him to see the possibilities and have the knowledge, skills and confidence to take on his life-long adventure with tea. It is no wonder that when the Lipton Tea Company abandoned their substation study of tea plants after the 1979 hurricane, Donnie saw the opportunity and collected the discarded plants which became the root stock for his tea plants.

Prior to retirement, Donnie had a long and successful career as a graphic artist, designer, and school teacher; he volunteered as an artist, archaeologist, historian, museum curator, Civil War reenactor, designer, event organizer, collector and speaker. His experiences, knowledge and talents culminated in Donnie becoming the recognized local historian in Fairhope.

In 2007, the City of Fairhope and the Fairhope Single Tax Corporation joined efforts to renovate and expand the old 1928 City Hall building for use as a history museum. The Fairhope Single Tax Corporation president said the project would happen only if Donnie Barrett was named as director. Donnie accepted the offer, and the Museum opened in 2008 showcasing Fairhope's history with his imaginative exhibits and creative displays. Hundreds of thousands visited the Museum during his tenure (2008-2018) learning about Fairhope's unique history through his exhibits, programs and events.

It is through Donnie's work as the founding director of the Fairhope Museum of History that many know him. The Museum's volunteer docents were trained by Donnie himself, who greeted and updated them about the Museum goings-on at the beginning of every shift. On his suggestion, the Friends of the Fairhope Museum of History was organized in 2010 as a non-profit group to raise funds and support the Museum. That led to Friends-sponsored fundraising events like festivals, Colony Cemetery tours and Historic Bus Tours – all with Donnie's leadership and participation. The weekly highlight was the Tea For Two – when Fairhope Tea Plantation hot tea was served to guests as they listened to a speaker present information about local history. Donnie's introduction of the speaker was preceded by a bit of information about tea, with him waving a branch broken from a tea plant growing outside the Museum.

Many bags of Fairhope Tea Plantation tea were sold at the Museum gift shop adding hundreds of dollars annually to the shop's revenue – all used for Museum projects. Donnie and Lottie donated tea to the gift shop, only asking that the Docents and Friends help pick and process tea at the farm. Many of us who were in the fields and the processing center (that was the Barrett's patio) by the backyard peacocks remember those fun afternoons and the delicious aroma of toasting tea leaves.

Before retiring from the Museum in 2018, Donnie remarked that tea was his and Lottie's retirement business. They planned to devote all their time to tea which had been a side business for many years. And that has certainly happened – in fact it is a full-time job with weekday tours (by appointment), and the seasonal cycle of picking, processing tea and maintaining the farm.

The Grandfather of American Tea explores one man's fascination with the global beverage of the masses and the science and craft of tea. It is written in Donnie's unique "elegant Fairhope folk style," and is a delight to read.

Finally, I'll mention that years ago when my family moved to Fairhope, one of my eighth-grade teachers was Catherine Barrett. Mrs. Barrett was a wonderful teacher, and she had a chic French twist, wore beautiful dresses and suits, and (this was special) our names were spelled alike. Once in a while she mentioned her twin sons to the class, and we perked up to hear the news. So, forty years later, it was a treat to finally meet one of those sons, Donald Hatchett Barrett, when the Museum opened and I became a Docent. Donnie is a friend to many, and for me, he is a brilliant, generous and gracious person who has kindly shared his knowledge of Fairhope's history. I am proud to know friend Donnie, who is, unlike Hurricane Frederic, a real gentleman.

Catherine King

Fairhope, Alabama
Fairhope Museum of History Docent
Fairhope Single Tax Corporation Archives

Introduction

My wife Lottie and I have been making tea for 46 years. It started out simple with my family making fun of me, Lottie saying "Oh, you and your tea," to all the family, to now being supportive and surprised that we have been so successful. It has been a bumpy ride with many more failures than successes. Finding out there was no guidebook to tea making led to years of experimentation which has never really ended.

When I recovered the first tea plants in 1979, I found out quickly that there was no one else growing or making tea, it was very exclusive. That uniqueness might have been the driving force to me learning how to make a good brew. We now have a successful "Mom and Pop" business.

As it was a hobby for about 25 years, it did not matter if my tea was fit to drink or not. Because I had no funds invested, I could play with the leaves in any fashion I wished, which is still a bit true today. I do not have to please anyone as I would if I had to attract and hold customers. I have never depended on tea for a living.

Donnie and Lottie Barrett, "Pop" and "Mom."

"The Grandfather of American Tea" is quite assuming and presumptuous. I was called that by a tea distributor over twenty years ago. There are over 100 tea makers in America now that I do not know, and they have never heard of me, but I believe my farm is older than most. As you "gotta be self-promoting and blow your own horn" in this world, I have used that name.

This work is more of a memoir, a journal, almost a diary of what I had done over the years. It is not a how to book but, a what I have done book. There are several books out there, such as the profound and well researched and expertly written "Grow Your Own Tea" by my friends Susan Walcott and Christine Parks.

Some of the dates as to when I did this or that may be conflicting as I did not keep records and have trouble remembering. I often have trouble remembering what I had for dinner last night but the dates are close.

After so many years of thinking about tea, studying it, communicating with the few others in the industry about it and the endless experiments, I thought I may need to record what I have done. As I seem to have no understudy lined up to take my place, this seemed to promote this work. I want to record my years of work with tea for anyone who may find it interesting.

Donnie Barrett

The Grandfather of American Tea: My Tea Story

Donnie Barrett

This is a photo I made of a Chinese tea field in either Wuhan or Nanjing, taken from a train window. The Chinese claim that tea originated in China which is partly true.

History of the Tea Plant

There are several accounts of where tea came from, who first used it and how it spread from culture to culture. When reading these there are several notable conflicts which I will leave to others to sort out.

It is generally believed that tea is indigenous to the foothills and ridges of the Himalayan Mountain range. A map on Wikipedia shows the north range is about where

Tibet and Nepal come together. To the west, in the Assam states of India a subspecies evolved on the western side of the mountain range. The indigenous area goes south, down through the Yunnan province of China to the top of Myanmar. The Chinese claim it originated in China but that's not completely true.

The stories of which Chinese Emperors found, discovered and promoted its use are historical folklore that likely changes from region to region and has evolved over the years. The stories of how leaves from a nearby bush fell accidentally into boiling water are amusing.

The British record many expeditions and exploitations of the lucrative tea business. They documented how the practice of tea enjoyment swept the entire world. Tea is now grown in many countries with their cultural practices attached. Tea is processed in so many ways throughout the world that is surprising it all comes from the same leaves.

The history of tea in America is said to begin with a French botanist who planted tea near Charleston, SC in 1799. Many other efforts from New York to California down to the Atlantic shores of Georgia have seen tea plantations come and go. Even in Mobile, Alabama, the British brought in thousands of tea plants when the area was British West Florida during the late 1700's. No tea farm ever developed there. There are now well over one hundred farms in the USA growing tea.

When I was in Tibet in 2007, it seems they drank mostly Jasmine tea. On all of our side trips to different locations to usually see another Buddha cut into the side of a mountain, I would walk off the trails out in the natural terrain. I think I saw tea in several places under heavy fir trees. My guides did not know any names of the plants, trees or types of rock, but I thought it was definitely tea.

Americans drink mostly black tea because Mr. Lipton bought most of Ceylon in 1906. The large island nation, now known as Sri Lanka had an economy based on coffee plantations. By this time, disease had killed the coffee bushes and the financial system and government were failing. Mr. Lipton started using the technique of oxidizing the leaves to make black tea. The Chinese claim oxidized tea was first produced in China but historians believe it was an Indian process.

Mr. Lipton owned the fields, the factories, the vessels to ship the tea and owned the racks on the store shelves, back in Briton, Canada and America, to sell it. World War ll sealed the trend, when tea could not be brought from the oriental countries. When our great grandmothers sipped the Lipton tea they highly enjoyed the robust flavor. Before that, everyone was drinking green tea. Even at the Boston tea party they threw bohea, a cheaper form of green tea blend, into the harbor. Of course, the Boston tea company who famously sells the "Black tea that the "Indian" patriots threw into the harbor" would completely disagree.

The history of the tea plant is a complicated and endless story.

How I got Started

A life changing moment doesn't always announce itself as such. That is what happened to me in the fall of 1979 when my father, superintendent of the Auburn University Agricultural Substation, and the legal manager of all the experimental station materials, showed me the destroyed Lipton tea plant experiment and said "Here, let's get you up some." I did not know a life changing moment was at hand. Life changing moments seldom announce themselves as such. My interest in the plants was mild at that point but with three broken up bushes in the back of my white Toyota truck, my interest in tea would sky rocket!

In the mid-1970's I had returned from Auburn University with a newborn little girl, and a job at Poser Business Forms. I thought I would work there a few months which turned out to be 31 years. I was living in town and came out to the Auburn Sub-station often. About 1976, the Lipton Tea Company planted an acre size experiment or a demonstration of perfectly aligned three-foot-high tea bushes. They were identified with Chinese names written with a black grease pencil on wooden stakes.

We marveled at the plants wondering how you get the amber juice out of those green plants. We had in our refrigerator a stainless-steel Army surplus pitcher (which I still use) with a shovel load of sugar in strongly brewed Lipton tea. We drank sweet tea at lunch and supper all year long, all our years growing up. We, like everyone else then and now, had no idea where tea comes from and new nothing of how it was made.

The Auburn Agricultural Substation, now known as the Auburn Research and Extension Center, was a great place to grow up. My parents lived there when I was born down on the Bay. My father was hired there as Assistant Superintendent right out of Auburn University in 1947. We had free building materials, tools and a space to produce any project that you could dream up. We had a large walk-in cooler which was always a cornucopia of fresh vegetables. They planted test plots of every sort of crop imaginable. We often saw plants that we had never seen. That is why we took for granted the tea plants. They did research on crops, animals, chemicals and equipment which put us around college professors on a regular basis. They called plants by their scientific names which I still do today. My brothers, Ronnie and Jimmy, also knew the names of most every plant, tree, bush or weed.

In September 1979, I was out there dumping hurricane Frederick debris in a designated Corps of Engineers dump site on the Substation. My father said, "Come look at what they did to the tea experiment." We saw they had run a bulldozer all over the four-foot-tall bushes, pushed them up in six-foot-high piles, poured diesel fuel on them and burned them. In 2020, a Mr. George Stillman came to visit us, and told me how he had planted the tea and later, about two and a half years later, bulldozed them.

Several of the piles were dug into (likely by a local nursery) and root balls and broken limbs were scattered all about. My father said "here, let's get you up some." I likely would not have picked up any tea plants except for his encouragement. We had been Camellia enthusiasts with grafting, collecting and rooting sasanqua as grafting stock. We knew how to look at the leaves and tell different varieties. There were four there, we did not know, so we got three broken up bushes and put them in my truck.

When I brought them home, they were naked root balls and broken limbs. I had rooted Camellia sasanquas for graft stock as a teenager and sold several to my father's friends for $1 each. Also, when I was at Auburn, I worked at Dr. Norton's nursery rooting scrap from landscape jobs to make nursery stock. So it was natural for me to start cutting the broken tea branches into rooting scions and putting them in wet sand in old wood printing plate boxes with window glass on top. There was no advice, guide book, manual or any instruction on what I was doing. I did not know that I was starting the first modern commercial tea farm in the United States.

The Grandfather of American Tea: My Tea Story

This is me lying in the sun at my parents house around 1982. I had hair…and it was black. Behind me, near my father's wood pile, are two of the original Lipton tea plants we recovered from the burn piles. They were planted there only three years before but the large root balls grew quickly.

After I had lots of tea plants growing, I found out that you could not simply dry out the leaves to make tea. It makes dirty water. Then I found out it was called "tea craft" and it was a secret process. No one was there to tell me how to make tea, and being before there were computers to use in research, I went to China in 1984. I went to several tea factories and tea farms all over China. I asked lots of questions like a simple tourist but I knew what I was seeing. I actually stole trade secrets from the Chinese and learned to make commercial grade black tea.

In this 1984 picture I looked like I was photo shopped into a 18th century photo, being totally out of place.

When I returned from China, I first made small batches of my poor-quality black tea and after I sorta learned how to make it, I would make small packages to sell. My first labels called it "Hurricane Blend" because I thought the hurricane in 1979 had swirled all the plants together and that was some type of blend. But it wasn't. It was years later, when I was back out at the Auburn Experimental Station doing a presentation for the Master Gardeners, who regularly met there. The superintendent there at that time, called me around the corner of the barn like a boy who had done something wrong and was a bit ugly. He said the tea experiment was not terminated by the hurricane as I had said, and that the Lipton Company had just decided to terminate the planting. He dramatically waved his arms in the air and wiggled his fingers and said "and, no one knows why!"

Within a few months I had over one hundred tea plants in the ground and in black plastic pots. That's when I learned three things about tea.

1) All types of tea come from the same plant.
2) If you dry the leaves out, it does not make tea. It makes dirty water.
3) Making tea is a secretive process. It's called a Family and Industrial secret.

Cold frame tea in pots

When I first started rooting the original tea plant cuttings, I used wooded boxes with plate glass on top. When I moved to Barnwell (Cavanac) in 1990 I built my first cold frame green house. This is a type of greenhouse where the cutting trays are placed on the ground during the winter, and it uses the heat of the Earth to stimulate root growth. I have no idea where I learned how to do this. It was a 2 x 4 frame with a plywood roof that would be completely enclosed at first and then completely open after the rooting process had started in the summer. It was built out in the open which was a challenge keeping sun out. I later built one twice as large, 35 x 15 ft. in a shaded area but the tea was prone to getting fungus. The cuttings would remain in the trays almost a year before being moved into pots or out in the fields.

*This is my first cold frame greenhouse built in 1991.
We rooted 500 plants in each tray.*

I first used wet sand and Rootone rooting hormone but later started buying large bags of vermiculite. This retained the moisture better. I did try Perilite but I think it was more expensive. I would use the vermiculite for several years.

This is me rooting tea. The scions are dipped in a rooting hormone and placed in rows in vermiculite. This produces a clone of the parent plant which carries the same DNA. The leaves are clipped to reduce the water-loss liability of the cutting and allow more to be placed in the tray.

I first planted clones into black plastic flower pots as well in my first tea rows "out by the pear trees." I thought I would use the potted plants to replace any casualties out in the field but that never worked. My potted plants transplanted in the fields would just die. I did start selling them for $25 each which were my first good sales. We had a "fly over tornado" in 2001 that topped a large Cedar tree and the large top crashed directly pointed downward in my potted tea garden which was the end of my tea potting phase.

The first rows of F1 seedlings did well before I started planting clones from the cold frame. That is where I developed my tea plant covers like I saw in China. The Chinese used 18 inch square bamboo boxes, open on both ends, when placed end to end would make long rows and rows across hill sides. Later I saw them using more modern shade cloth and fiberglass rods. I made covers of cloth, black plastic, plywood and roofing tin for covers. Some worked better than others. I lost lots of plants before I started developing a serious cover. I later graduated to silt fences, as you see around construction sites, augmented with black plastic.

These are the covers made from silt fences.
I tied them forward to protect from high winds.

This is a tea cover like the ones I first saw in China. They also were using little bamboo boxes which hooked together in long rows. In 2007 I saw them using nylon rods bent over with a shade cloth covering.

Two lots

In 1979 when I first found the three broken tea bushes I immediately rooted over a hundred of them. I started planting them in my parent's back yard across the two empty lots they owned next to their house which they had just built in 1978. It was a very open pine savannah with little under growth and completely unused for anything else. Over a several year period I had several rows across this area with hundreds of plants. This was where I first learned to make tea from the Chinese in 1984 and where I first sold a little not really good tea here and there.

By the late 1980's I had a substantial stand of hundreds of tea bushes at my parent's house. Lottie and I wanted to buy a house and I wanted a country farm so we moved into a trailer park for a couple years to save money. We found a house with 5 acres on Lyter Lane in 1989. This was to be the second location for my tea farm. We bought this farm for $49,000. We finished the remodeling and moved there in January,1991.

A photo of the Washington Ave. two empty lots garden, taken in 2000, showed the tea bushes grown up tall after ten years after I moved the tea operation to the Lyter Lane farm. You can see in the photo that the open pine woods with little underbrush had grown up thick and heavily wooded. Also, you can see that my parents had started to use the area as their back yard. There is now landscape timbers along their drive way, bird feeders and a useless bird bath. On the left is one of the fifty foot rows where my father soon would put a utility building right on top of the bushes, then another and then another to destroy all the tea plants there. He thought I had no more interest or use for the tea plants growing there. This yard now belongs to my daughter and her husband. I now use this area as a nursery for growing seedlings.

The tea plants on both sides of this photo are the original plants from the Lipton experiment and rooted clones I made from those plants.

Donnie Barrett

Growing Tea

The tea plant is classified as a camellia. Basic needs for a camellia are lots of water, acidic soil, some understory shade and well drained areas. That sounds simple enough. Most of us have a camellia bush in our yards which requires no care, and you seldom notice it except when it blooms.

Tea has those basic needs but getting tea started and established can be tricky. I sell thousands of plants a year and I would say only half of my plant customers are successful. That is mostly because they are experienced gardeners and don't completely follow my advice. Growing tea is forestry and not gardening.

Most folks who want to get a nice plot of tea going fantasize digging a small, shallow trench, laying down a row of seeds, squirting it with the hose and poof you have a row of tea. It is not that simple. I'll give away seeds every year but don't sell them, and have not tried to establish tea by using seeds. But I have used seedlings and mostly cloned rootings.

The seedlings I have used and sell are the first generation from the original plants I recovered from the Lipton tea experiment in 1979. This is known in the genetics world as the F1 generation. Because tea is heterozygous you do not want to use seedlings from other seedlings, from other seedlings. You will breed away the high quality traits that make good tea. I knew about this because of my college classes in genetics and have been complimented as likely having the only intact cultivar in the US, which I call Camellia sinensis *fairhope*. There are four (or five) universities growing what they call Fairhope Select. The University of Mississippi reported that Fairhope Select is outperforming the other cultivars.

Tea seedlings from under the original Lipton bushes. Geneticist refer this as F1.

What I have used the most of, to establish my approximately 61, 000 bushes, were rooted cuttings. First that gave me lots of clones with the same genetics as the original Lipton plants. My first tea farm, the two empty lots my parents owned is most all clones of the original three plants. When I sell plants, usually 3-5000 a year, I select the seedlings from there where my daughter, her husband and my grandson now live. I've not rooted cuttings in 20 or so years.

When I observed the Chinese tea farms, among many other things I noted was the spacing and their methods of shading the young plants. I adapted what I saw into my own tried and not so much proven techniques.

The Chinese planted the young plants very close together. Crowding them was said to make them try harder. So, when I planted my rows I would dibble a hole, put three rootings in and dibble another hole beside that hole and plant three rootings in that hole. I would come forward 10 inches and repeat with two holes side by side. This crowded the plants very much, 1000 young plants in a 50 ft row. This is how the Chinese were doing it.

Modern consultants now days are recommending spacing young plants as far as two and three feet apart. These rows grow up thin and airy and not hedge like. I am not sure if the recommendations are for better plant life or health or if they are selling the customer expensive young plants and want to spread out the planting so the customer will be more comfortable. I have found the Chinese technique has worked very well for us here in Fairhope and for the others I have advised.

When I first went to China in 1984, they were shading their new plantings with little bamboo boxes. They were open on the bottom and open on both ends. They would place these end to end over the young plants and this produced long rows in parallel order snaking over the sides of hills. The bamboo boxes had loose slats in the top that could be removed slowly to allow the young plants to get used to direct sun light. After a year and a half the long rows of bamboo boxes would be removed and used again on fresh plantings. I first thought that is a lot of trouble but soon learned with practical experience that it was necessary, and that you were really training the shade loving plants to survive in the sun. I also learned that shade grown tea, in its natural environment, was not as good as sun grown tea.

For shade I tried many techniques. My first attempt was a roll of barbed wire covered with scraps of cloth to deter chickens. My aunt was a quilter and she had left me bags of scrap cloth. I learned white cloth was not cover enough whereas colored material was dark enough. I started making racks on wooden stakes with first cardboard covers, then nylon material (the loose threads would bind and strip the young plants below) then black plastic. I made many versions of these covers with plywood, roofing tin and feed sacks all in the Chinese bamboo style. As stated, the best covers that I adapted were made with silt fencing, as used on construction sites, leaning over the young plants. I would staple splices and tucks and stake and tie forward for storm resistance. My many customers have come up with brilliant ways to cover baby tea plants.

Silt fences used to shade the young tea plants.

I ran hoses from my pump house out into the field. With the little plants, I would leave a depression in the rows to facilitate watering by holding water and allowing it to run down the depression in the rows to water the others. I once installed a soaker hose system that never worked well. The water line was useful for a few years while I was planting but I removed it as I never water the mature tea plants.

When I was in college, I was a hippie who thought chemical fertilizer was poisonous. I still kinda do and use very little fertilizer here on the farm. I have gone as long as eleven years without putting any on my tea rows. When I cut the old rows all the way back to a few inches from the ground, I will fertilize those with 8-8-8. Soil sample tesst recommend a 0-0-8, potash fertilizer. The young plants do not like any fertilizer and will develop leopard spots with even a light application, as my customers keep pointing out. "Why am I getting these spots on the leaves?" The best fertilizer that I have used for all these years is a truck load of leaves, its natural understory plant food.

During my first years on the two empty lots I did put most of my rootings in the ground but I also put lots and lots of them in plastic pots. Knowing I would likely be moving on from my parent's large yard, I had produced rows and rows of potted tea plants. I also thought they would be good to plant out in a field which got me started at the Lyter farm site, but the pots were not much useful after that. A 2001 overhead tornado tore off a large cedar tree and crashed it downward in my tea pot garden and effectively put me out of the tea in flowerpots business.

After getting several rows of tea plants established on my Lyter farm location, I had reached a satisfied point where I had plots resembling a commercial tea farm but it was mostly still just a hobby. During the 1990's I made small batches of tea, had little Ziploc containers to sell, but I saw no need to keep any kind of records as it was a mere hobby. I started keeping records in 2004 when I produced 77 ½ pounds of tea and made $1570 in sales.

Pruning

Pruning is a winter time activity we do every year. Most tea producers pluck most all the growth from their bushes and do not do aggressive pruning. Our tea bushes outgrow my ability to pick them and continue to grow in the fall after we have stopped picking for the year. Therefore, they get a good "hair cut" in the winter before the spring flush.

We never pruned this acre of tea that we planted at Lottie's parents home and it grew twenty feet high.

These rows are freshly pruned while the rows in the background are not yet pruned.

Pruning is essential to produce a plant with a strong frame to support lots of little branches that form a picking table. This starts when the plants are very young. You want to produce a bush and not a tree so you first pinch the central growing stem down to about six inches. You want the plant to fork and branch out.

I keep this branch on my back porch to use as a tool when instructing new growers how they need to encourage branching.

This is very difficult for new growers to do. Tea plants grow in stages, one person described it as "sleep, creep then leap." When one of my tea plant customers finally sees the little plants, they have been babying shoot up tall, it is hard to get them to cut them back down to six inches. But that is what has to be done to make a tea producing bush.

After a few years of pruning to encourage branching, depending if you have strong or weak plants, you will begin pruning the bush into a prescribed shape. Traditional shapes are called "cupcake" or "wineglass." I still have several rows shaped in this style but they prove to be harder to pick. More modern tea producers are most all going to a flat, "pancake" shaped hedge row, especially if picking machinery is used.

I prune some rows flat and some round top.

For many years I would use the traditional round top rows to grow an older pluck for black tea, picking about once a month during the picking season. The older, taller growth was just easier to pick off the round top row. I liked this older pluck because I thought it made a stronger black tea.

I would then use the flat top rows to grow the picking for green tea, picking about every two weeks. This produced a younger pluck, which I think has more sugars and starches for making green tea. I'll have to admit that later in the picking season I do not stick to this and walk around the rows picking the age of leaf I have in mind.

After several years I will do a "hard prune" on a row and cut it back down to about six inches. The rows recover quickly, producing a strong growth of juicy buds. Some foreign producers cut their bushes down like this every year because of the high-quality tea produced.

The row on the left was hard pruned.

These hard pruned rows will appear wider and have better branching and produce more little stems in the picking table. They just seem to try harder and produce more tea to pick. I like the effect of this pruning technique but find it is a lot of work and can only do two or three rows a year.

Another style of pruning is a "skiff prune." This is applied to an overgrown row during the picking season if the bushes have new growth that has gotten too old to make good tea. Because tea grows so unevenly, and out grows my ability to keep up, this happens to me every year. But, after trying a skiff pruning for several years even with side-by-side comparisons, I find I am pruning off the tiny developing buds of the next flush. This will cost me a crop of tea whereas I find it better just to pick around the older growth.

Tea Seeds

Lots of discussion has been made over the use of tea seeds to establish a stand of tea plants. Less than half of my plants out in the fields are grown from seedlings but most were plantings of rooted cuttings. These clones retain the same genetic properties as their parent plant where as the seedlings are of a slightly different DNA makeup.

Some growers and tea producers think rooting cuttings is the best way to establish a tea crop. I was first introduced to this line of thinking years ago and rather still have this frame of thought. Tea is heterozygous like us humans, we don't look like our parents and our children don't look like us. It's the way the flower has to be fertilized by another plant's pollen to make a uniquely new genetic plant.
That is why if you grow a Camellia *japonica* from a seed, it's your own plant and you can name and copyright it.

Same is true for apples. Johnny Appleseed was spreading new varieties as he spread the apple culture.

This is good for a diversity of flowers but it is not good for tea because when heterozygous gametes are formed, the characteristic traits of the two parents sort out. The traits of the tea plant parents are distributed unequally so some have this quality and some have that. What happens is after several generations you lose the robust and aromatic qualities to make good tea.

A good example of this is most tea plants from Home Depot, Lowes or a community nursery do not make good tea. The desirable aromatic qualities have been bred out. The tea is flat and uninteresting. Another example is when Mobile was British West Florida (1761-1784) they knew tea would grow in this climate so they imported lots of tea plants to Mobile. The fine citizens of Mobile never established any tea farms but planted the bushes around their homes. Now years later, descendants of these bushes are all around the upscale homes south of Point Clear, only a few miles away. Every year, someone will bring us a trunk load of yard clean up debris and want us to make tea. I tried several times, and it makes flat, uninterested dirty water that looks and smells like tea. The good flavors have been genetically lost.

It takes a year for the seed pods to mature.

    Many tea producers are inclined to be against tea clones from rooted cuttings. It is true that a cutting does not develop a tap root but they do produce large root balls. If you dig up a mature plant, which is nearly impossible to do, you will notice the massive root ball does not have a tap root. I have also read where foreign tea producers say cuttings do not have long lives and waste away quickly. This is questionable. One tea farmer told me that you can always tell rooted clones because they were always "yellow and unhealthy." I think this too is untrue. Others say you need the genetic diversity of seedlings and have made up plans for planting plants of different cultivars in plots to enhance seeds with a mixed genetic diversity. They also admit, "you don't know what you are going to get," when you buy seeds from someone.

    One of my plant customers bought 500 seeds for $250 from another tea producer and sprouted just 20 plants. He came to me and I gave him a large box of hundreds of free plants and he will always be my friend.

Other tea producing friends are adamant that buying and establishing a tea garden can only effectively be done using seeds. People thinking of growing tea often think if you dig a long shallow trench, place a seed every foot, cover and squirt with a hose a row of tea will pop up. This works with corn but not with tea. Our cold frame green house was very effective at producing thousands and thousands of tea plants which have been very productive and we are still harvesting here thirty years later.

After a quarter century and thirteen hurricanes later, part of our cold frame still stands.

Several years ago, I was picking up seeds and selling them to a cosmetic factory in Birmingham, AL. They would crush the seeds, boil them and extract "tea seed oil" after it cooled. They were using this, and advertising such in several cosmetics. I did not like collecting the seeds and recommended them to another tea producer and lost contact with them.

I have another regular customer who comes every year around November before the tea pods open and picks large buckets of green seed pods off the bushes to dry and sprout. I give them to him for free. He seems quite pleased with his results and is building a large garden of plants.

Plantation?

When I first started selling tea at the Fairhope Farmers Market in the mid 1990's I needed a name and label for my little Ziplock cartons of poorly made black tea. I called it "Hurricane Blend" because I thought the 1979 Hurricane Frederick had mixed up all the rows and lost the name tags, wooden stakes with grease pencil names, on the rows. I also thought that the hurricane destroyed the experiment and that was the reason the field was bulldozed. I found out later that it was not the case. I made a rubber stamp (I made these at Poser's) and stamped the name "Hurricane Blend" on the carton lid.

LYTERSIDE TEA PLANTATION
MAKERS OF *HURRICANE BLEND*
FAIRHOPE, ALABAMA

Later I called it "Fairhope Tea Company" and made labels with "Hurricane Blend" in a smaller circle on the label. I made lots of these and used for several years, stamping on paper and glued to different boxes, fruit jars and bags. The name tea company did not indicate that the tea was produced in Fairhope but likely was an import.

When I started selling the donated tea at the Fairhope Museum, I simply called it "Fairhope Tea." Museum gift shop customers liked that it had "Fairhope" in big letters across the label. We sold thousands of dollars of tea, mostly Oolong style, there under this name with proceeds going to the Fairhope Museum. Because of ethics considerations, I did not make one cent selling tea for eleven years.

It was about 2009 when a lady at one of our weekly "Tea for Two" programs said, "I'll bet you have a plantation!" Good idea lady, I'll start calling my company the "Fairhope Tea Plantation." The name "Company" insinuates that I buy the tea from elsewhere and resell it. Plantation tells folks I do indeed grow the tea. I had forgotten Charleston was known as a plantation and would have not gone in that direction, but I did.

This has caused a few problems. First customers would drive up expecting to see large, white Greek Revival columns on a long circle drive. One car full of ladies asked if we had any slave quarters. Many, many cars have driven up then drive out when they did not see what they thought was a plantation home.

Unfortunately, when guests drive up to our farm they come up to our back door. This isn't exactly a "Tara" plantation home. I do plan this year to build a kiosk there by my truck with a large sign, brochures and teabags to sell.

The Webster definition of a plantation is: "a planned estate on which crops are planted and cultivated by resident labor." We bought this farm in 1989 to grow tea and my wife and I are the only labors who do indeed live on site. People do not know this and think of a plantation as in the Gone with the Wind movie.

I have also had some really negative comments connecting the word plantation with slavery. We obviously have nothing to do with slavery but people will bring it up. Often it is news reporters or magazine writers that ask me about the suspiciously leading name. And they sharpen their chins when I explain our intent, thinking of me as a Klu Klux Proud Boy. Our worst insult was the church camp just three miles east of us who regularly brought groups out for a cup of tea and farm tour. They all-of-a-sudden decided that they could not come here anymore to our "evil plantation." I think I was not even charging them so it is their loss.

I am now getting where I call this the Fairhope Tea Farm because it is easier to write and does indicate that we grow the tea. I would and likely will change over to the "farm" name, it is just that I now have so many labels and signs with "Plantation."

Learning to Make Black Tea

All the time I was growing up at home on the Auburn Experiment Station, we kept a stainless steel, army surplus pitcher in our refrigerator. This contained strongly brewed Lipton tea with a shovel load of sugar. Sweet tea! It was a Southern staple and we drank it all year long. This was what tea was. There were no other "teas" out there. Lipton tea was the only tea. The terms "black" and "green" are very modern terms. Over the past centuries, tea was called by the factory that made it, i.e. Bohea.

I still use my mother's army surplus pitcher.

So, during the early 1980's when my tea plants were taking off, I tried to make tea. Of course, I picked young leaves and dried them out. That didn't work so I started experimenting with little batches knowing as little as I knew I had to do something to make my "tea" taste like Lipton tea.

My first slight success came when I picked young leaves, beat them up, and mashed them in a shallow bowl to sit there a day or two before spreading out in the sun. I had read in a general magazine article that tea was "fermented." I had made wine so I sorta knew what that meant. This put a light brown color to the brew, and it mildly tasted like tea. My first success, which was almost accidental, made a watered down, weak brownish cup of tea.

This early experimentation was at my parents' house in the early 1980's before I went to China in 1984. I was in Dr. Jones dentist's office waiting room about this time reading a beat up, old National Geographic magazine. In an article on China there were pictures of Chinese laborers working out in a tea field. That was an "ah-ha" moment when I decided I needed to go to China to learn how to make tea. I announced this to my wife Lottie who quickly said she was not going so I booked the trip for just myself.

Donnie in China in 1984 looking like a foolish Maygoran (American) playing with Chinese children.

This was June, 1984 when I found myself going to China. Still, I had not tasted green tea and was still thinking Lipton tea was all there was. At the first hotel I checked into, they handed me two, hot water glasses of green tea. I thought it oddly weak, a bit clearish and not amber brown. I learned then that there were more types of tea in the world besides Lipton. It was a custom then, seems not anymore, to welcome guests with two classes of hot green tea. Every hotel, tourist spots and train rides were complimented with green tea. I immediately developed a liking for the unsweet green tea.

My first days in China were busy with several tour guides who spoke terribly broken up English. Tourism was new for them and I'm sure they thought they knew enough English to entertain tourists. We also had plenty of options for where we would like to visit or what we wanted to go see. I then started going to tea fields and factories which began my real education into tea craft.

Gathering that different teas were processed different ways, and different teas were in piles around each other, I started sorting out the different processes. I asked lots of questions and they answered me.

This is where I saw tea being oxidized. I am sure I saw tea rollers in action but I really do not remember them because I did not know about them. But, what I did see were large piles of tea under heavy canvas covers which had a wonderful smell when the canvas was lifted. They explained in detail what was happening and still being naive on the subject, I would only grasp part of what they were saying.

Tourism was new to China, these ladies had never seen westerners.

One feature that was common at most of the factories was long buildings, looking like a commercial chicken house, that were "withering sheds." They were long and narrow and on one end would have fans and heaters to blow hot air through the length of the shed. There would be shelves, only inches between them, with thin layers of tea in flat baskets on each shelf having warm air blown through them.

I tried to mimic this process in several ways when I got back home but had little success. I remember an older Chinese gentleman tour guide saying "solar withering" so I used that technique for several years. Actually, learning many years later that cool withering works much better.

I saw the Chinese drying tea in several different ways. Seems like big hot ovens were in all the facilities we visited. I really do not remember any discussions of drying in ovens but I'm sure there was some. This is about the time I saw tea being dried in little rooms with ceramic or brick walls only a few feet high with no roof. This is how I learned to use the dry pen technique I used for years and still use.

My first dry pens were stakes in the ground with old tarpaulins or plastic stapled to the stakes. Every passing storm would blow them into the woods.

One small village we passed through would block off the paved road (their only paved road passing through town) and sweep the pavement and dry their tea on the black top surface, which got hot in the sun. They would direct traffic (there were very few cars then) to drive around on the downwind side so dust would blow away from the drying tea.

This basic exposure to tea craft put me on the right road to making black tea. When I came back from that first trip I could make tea that tasted a lot like my mother's Lipton tea. I thought I had arrived. By the late 1980's I started making little batches of black tea, packing a double hand full in a Ziplock plastic bowl, and selling these at the

Fairhope curb market (on Church Street by the "Tomato Lady") under my first product name "Hurricane Blend."

I called it that because the original plants from the Lipton experiment were mixed up by Hurricane Frederick, Sept 12, 1979. They had Chinese cultivar names on the different rows but all that was blown away. I recovered three different varieties but did not know their names. I also thought the experiment was terminated because of the hurricane but later learned that was not the case.

By the end of the 1980's my tea farm on my parent's two empty lots was getting too small. Also, my parents were spreading out their yard and my tea bushes were clashing with their patio and garden endeavors. I told my lifelong realtor friend, Bobby Mannich, that I wanted to buy a farm.

Towards the end of 1989 Bobby called and said they are selling a brick house on five acres of land for $49,000. "Let's go out and see what the problem is" he said. He picked me up at lunch and we drove to the end of Lyter Lane and saw that the property was a mess.

The last quarter of Lyter Ln was plowed away into the fields every year and sometimes a jeep was needed to get to my house. The property owners had been throwing their garbage into the woods leaving a beaver dam wall of garbage around the back yard. The fireplace was bricked up, the appliances stripped or gone and most all the electrical receptacles were removed and taped over. But I was home, I never left and after a yearlong renovation, this became our tea farm in January, 1991.

I was getting tired of selling my little Ziplock tea packets at the Fairhope curb market. I was exhausted every time explaining tea over and over. The guy next to me selling cucumbers did not have to explain cucumbers. I stopped going to the curb markets even though I was making $3-400 each time.

During this time, I called my business "Lyterside Tea Company" until I realized it indicated my tea was on the lighter side, which it wasn't. I was selling pounds of tea to a business in New Hampshire called "Coffee and Teas of Yesteryear." They thought my tea was good for Chinese style fruit popsicles. This attracted tea distributors who came down to Alabama to see me and buy my tea.

One of these gentlemen, a Mr. Marshall, was not complimentary at all of my black tea, the only tea I made. We were grinding it up fine due to Shen Hua's suggestion which was not good advice and Mr. Marshall called it stale. But he lent me a book for a year called "Twenty-Five Years of Research at the Sri Lankan Tea Institute."

The book was large and heavy and contained a large number of scientific research report papers. It was not a narrative of their research, just a collection of reports from experiments and demonstrations. Being educated in scientific vernacular I enjoyed reading them. They covered many subjects such as how close or crowded to plant tea plants, how to prune, when to prune and different pruning techniques. There were papers on the tea craft, oxidizing, withering, drying and results with sugar, starch, essential oil content and lots on different flavonoids. There were detailed charts of how one chemical reaction led to another chemical and experiments with the bacteria involved in the tea making process.

As I read these scientific papers, I really did not realize how over my head all this really was. I was just learning how to simply make tea and this was research conducted by experts, which were actually the finest in the field. I would read which leaf to pick for the different teas and I thought "I'd never do that" and how to prune tea hard and I thought "I'd never do that," and how to feed and water tea. I just thought I would not be doing any of that and now after having evolved so much over so many years,...I'm now doing all of that! This Sri Lankan book was very helpful over the years and information I learned from it kept me on the right track in my journey with tea craft. It was this volume and what I saw in China in 1984, that formed my formal education and foundation on which I built my tea making skills.

I wish I had kept notes over the years as I tried one technique then another. Having grown up around research scientists made me naturally experiment, constantly. Every batch I made was using a new idea or a different style. I have gone in every direction with tea experimentation which made some batches for the compost heap but also got me where I am today making some pretty good tea.

Getting the tea to oxidize well and controlling the bacterial action was the biggest challenge. If my brew was not oxidizing well, I would let it stay smashed in a container for longer, which could cause a sour flavor of too much bacterial activity. So, I worked on this for years. I knew to get it warm so I would use hot water bottles, or put in on the hot water heater, or put it out in the sun.

We had a visit from a lady who stayed here for a couple days that was a British educated "certified tea taster." She was experimenting with making tea and we had some productive discussions. She was oxidizing her tea rolled up in damp towels she called dragon eggs. I modified this technique with old bed sheets to oxidize my black tea for a couple years. It proved to have mixed results, with uneven oxidation.

The dragon eggs would build up heat.

At this time, I was still working on perfecting my black tea. Sometimes it worked and sometimes not. Even though I saw lots of green tea being produced in China, I was not mimicking those techniques. If a batch did not oxidize well, I would called it "green tea," and sold it as such. I now would throw it in the compost heap. I still had a lot to learn.

In 1991 I moved my tea farm to the Lyter Ln location. I quickly built a cold frame green house and moved from my previous location several hundred potted tea plants. I thought I could transfer plants from the pots into the field. Also, I could fill a bare spot in a row with a potted plant. None of that worked. The first couple rows I planted did ok but after that all my transfers would die. I eventually gave away all my potted tea plants and did not use pots after that.

My tea craft techniques did not change much during the early 1990's. I was more focused on rooting cuttings and planting thousands of clone rootings in the field. During the mid '90's we planted thousands of rootings on the Jones land where my wife, her brother and her parents lived just east of where Lipton planted the test plot. In fact, you could see the plot from there. I tried to interest them in the project but they would not even water them. Same thing happened on my stepson's land. We planted lots of plants there too that just withered and most died. Then they decided to burn the land over, tea cannot survive fire.

By the year 2000 we tried again to replant the Jones field. It was successful this time. The plants grew into large bushes but it was so far from home that I only made tea there once to serve at the Jones' Thanksgiving dinner – just to make the point that this really was tea. There still is a half-acre block of tea there that is all so overgrown you cannot see it.

I started thinking about green tea then and was running green tea batches. But black tea was my interest as that seems to be what people wanted.

Around 1995 a Mr. Spencer Johnson asked me, at the old Fairhope Museum, if he could sell my tea at the Church Mouse, a British themed gift shop. They were on Section Street before they built the three-story brown stone building on Church Street. They sold tea pots, cups, trays, gifts of all sorts and had a nice selection of teas from around the world. They would make a pot of tea every day to serve customers. Spencer liked my tea and promoted it. I sold him a two-pound coffee can of my tea for $30. He showed me his books one day and half his sales were my tea. He sold many pounds of it. He carried my tea until 2007 when I started working for the City of Fairhope and had to stop all tea sales.

Mayor Tim Kant and Sherry Sullivan wrote me a threatening letter which I had to sign saying I would be terminated if I even said the word "tea" to museum docents or guests. For eleven years I did not sell one speck of tea. What I did was simply give most all the tea I produced to the Friends of the Fairhope Museum. They served it at our weekly "Tea for Two" speaker programs and sold it in the Museum gift shop. The museum staff would help make the tea they sold which was mostly

Oolong. We would have big picking parties and a week later have a toasting and packing party. Most folks who came were not good pickers but enjoyed the beer, chips and friendship. Lottie and I decided we could make the same amount of tea ourselves with the energy and expense of producing two social events for Museum docents. When I left the Museum in 2017, my tea made the Museum $22,800 in gift shop tea sales and much more with the two dollar gate fee for the Tea of Two programs.

Museum docents picking tea for the gift shop.

Withering the tea so it would oxidize was an important step that received lots of my attention. It sets up the phenyl esterase to do its reaction with sugars and starches in an oxidation-reduction reaction. But, it needed to be withered down to 55% moisture, which often was a guess. That is where the esterase works best.

Traditional Chinese withering techniques used heat. I tried electric heaters and small fans to simulate the "withering sheds" which produced a lack luster brew. I knew my tea was not oxidizing well and was not as strong and bold as I thought it needed to be. I withered mostly in the sun, "solar withering," which was just ok.

My best break came when my good friend in Tallahassee, FL, Mike Loeb, gave me good advice. He did much more research online than I did, and sent me a published article with graphs showing the chemical reaction durations and effectiveness over a period of time. This gave me insight into having my black tea develop a much better flavor. Mike also found and sent me a Youtube video of how to load and use a tea roller. These tips I am still using.

That's the story on how I learned to make black tea. What I am doing now is a long-evolved process of trials, experimentation, accidental discoveries and tips from other tea makers, a truly evolutionary process.

Learning to Make Green Tea

I did not know green tea existed until I went to China in 1984. I had never heard the terms "black tea" or "green tea." I grew a fondness for green tea during that visit and as I matured in my understanding of tea, that there were lots of green tea fans out there.

When I first went to China, I saw green tea processing but really didn't know what I was seeing. I saw several big (seems like brass) woks mounted in a brick table with a small fire underneath at a monastery. At factories, I saw large flat baskets being turned over large vats of boiling water where green tea was being steamed. I was a bit more focused on the black tea processing and rather dismissed the green tea processing. It was really two or three years before I started putting together a green tea processing plan. Before that, I called batches of black tea that did not oxidize well "green." It actually was more like Oolong, but I did not know that either. I now know the starting of the oxidation would destroy the natural fruitiness and lack of complete oxidizing would produce sawdust.

When I was lent the book "Twenty-five Years of Research at the Sri Lank an Tea Institute, about 1992, I read enough about green tea processing to start making real green tea. This book was a collection of scientific research papers and not a how-to directive, so I was still having to piece it all together.

My first efforts were a flat, hardware cloth basket which looked like a fireplace popcorn popper. The two halves opened apart where I would put several double handfuls of fresh tea leaves that I would roll round and round over my outside, propane shrimp cooker. I would get the tub of water rolling hot and steamed the tea. This produced an uneven steamed batch of leaves, but it made green tea.

At other times I tried heat treating the green leaves but lightly baking them. That kept the tea green, but I thought the fresh flavor of the tea was being lost by toasting it in the oven, so I only did that for a little while.

I cannot remember what turned me toward pan frying the green tea. Somewhere I read that the best green tea was pan fried and I remembered what I had seen in China and quickly put two and two together. So for twenty years or so, we heat treat green tea by pan frying it in an electric wok or large electric fry pan. We have gone through several, liking some more than others.

My wife Lottie likes the electric pans.

It seems that each of our batches is an experiment. Growing up on an experimental research farm and my training in college labs has had us try lots of different things. Also, with no "right way to do it" manual we keep on tweaking our processing techniques. We have gone through periods of getting the tea really hot and drying it a bit or just getting up to around 170 degrees. The leaf will be a bit limp with a bit of a graying color to the green. This is to deactivate the esterase or denature it to stop the oxidation reduction reaction which develops the black tea flavor. It has a preserving effect on the plant cells and metabolites.

Another effect we regularly do is after rolling the tea, we put it right out in the sun. Here it will oxidize a little in the hot sun. I call it "sun tanning" the green tea. It happens only where the sun heats it up the exposed leaf and depends on how much I rake it around as to how much it oxidizes. This gives the green tea a little boldness, a little more rich flavor. If you wait a day, either on purpose or from bad weather, this "sun tanning" effect will not happen. When making several batches of green, I will actually suntan a batch or two and not a batch or two and blend them together. We think this is our best green tea.

Cup of Chemicals

In college I minored in Chemistry. I took lots of biochemistry and nutrition classes. I enjoy knowing about the many compounds and their byproducts which are produced during the tea processing. This is a subject I enjoy talking about with the other tea makers I know. We seldom agree on much but talking it over is satisfying with someone who thinks this is as important as I do.

When I worked in a photo dark room, I had a chart on the wall that listed the chemical compounds that were found in coffee. It was startling to read how many chemicals are found in coffee. The same is true for a cup of tea. If you follow the line of compounds in the leaf developing into the compounds in the cup, it is quite overwhelming!

There is lots of research and published documentation concerning tea compounds and how they are effected by oxidation and toasting. I am not going to repeat all that here as most folks do not understand chemical reactions nor do they care. A simple search on the computer will produce an overwhelming amount of information on the subject. I will discuss a few of my favorites.

The many polyphenols in tea produce the flavonoids (or flavanols) which gives tea its complicated flavor. There are over a hundred of them in a single cup of tea. Varying amounts of them, produced are diminished during the picking, withering, processing and drying tea makes for the endless variety of the many different teas. Some relax you while others pick you up while others attack the free radicals in your body which many lead to cell damage and ageing.

The antioxidants in tea are the healthy compounds that tea is known for. The lucoanthocyanidins are the antioxidant compounds found in blueberries and elderberries which denote these as "health food." Another group are the catechins. One in particular is the EGCG, epigallocatechin gallate. This is a unique plant compound thought to reduce inflammation, aid in weight loss, lower blood pressure and prevent many common conditions like heart disease. It is found in many green vegetables. In tea, this is the strong grassy flavor in some green teas. Some makers, who do not like the "harsh with a single note" green tea, will oxidize their green tea to remove this flavonoid. This will produce an earthy flavor, which some folks like. I tell my tourists that green tea is either earthy or grassy, depending on the processing.

Another is the L-theanine. This is the amino acid that relaxes you even when you just think about having some tea. Modern science tells us it is produced in the stem between the second and third leaves and not found so much in the leaves. So when a tea maker holds up a stem says "we only use the bud and two leaves" (the video shows them picking magnolia leaves) they are holding the stem by the L-theanine factory.

Caffeine is most people's favorite polyphenol. It is classified as an alkaloid compound produced by the plant to deter herbivores from eating the leaves. Most people drink tea for this compound and for it's wake-you-up morning effect. There is conflicting information on whether the caffeine is affected by the processing or whether it is found in the younger or older leaves of the plant. The Japanese boast higher caffeine in their green teas but others refute that. I'll let the others decide this.

This is an extensive subject of which I've only made a tease. I have oversimplified this discussion to keep it light and airy and invite you to research the subject to answer any questions.

Plucking

Tea is made from the top leaves on the tea plant, the newer growth. That's easy but there is a world of chatter on when, how and how-low-can-you-go when picking tea. Tea picking is an art. It requires knowledge of the entire process and a snap judgment with every pluck. Many times, people want to come "help me pick." Sounds easy, right? It never works. When we had large picking parties for the museum Friends, only a few would be able to do it. Some would get a few handfuls in an hour and most would lie on the grass talking or drinking beer and eating potato chips in the shade. Tea picking is a journeyman level skill.

Lottie says I am the happiest picking tea. It is quiet with just me and nature.

There are several brands of tea that advertise that it is only picked in the morning dew. Is there a reason for that or is it a commercial ploy? They think it is important, but it would take years of experiments to answer this question.

There are descriptions out there of a "skiff pluck," "lite pluck" and a "deep pluck." If you have only a few bushes then such consideration can be given. This also goes with the serious discussion on whether one should pick the bud and one, two or three leaves per stem. Names are given for the different leaves, such as peco, cha, standard, rough, fish, mother and bangee, with charts with names and arrows. Companies proudly proclaim they only use the "bud and two leaves" to make their tea.

This tea folklore is good for publicity, but if you look at their happy pickers out in the field, they are picking the large size leaves. If your goal is to make money, and most tea is grown commercially to make money, you will be plucking older leaves. Sure, there are many teas, from all over the world, silver needle style and white teas that are just the bud and a leaf or two, but these are rather expensive and selected teas, not the bread and butter of the tea company.

We pick our tea as a farm crop. You pick the corn when it is ready and you can't wait too long or it will get old. This is how we look at the tea growth. I tend to pick the younger growth for green tea because of higher sugar/starch content and a bit older growth for black tea, which has more sticks, caffeine and stronger flavors. I tend to cut the flat rows (pan cake) for green and pick every two weeks and use the traditional round top rows (cup cake) for black, picking every three-four weeks. When this gets out of control (it grows faster than I can get to it) I move around the rows and pick the age of growth for the tea I have planned to make. When picking my hands move fast, grabbing a handful at a time and not stopping if I think I picked off a bit of weed, which is easy to remove during processing. I appear as non-discerning as a cow eating while picking and move along quickly.

One of the universities growing my Fairhope cultivar is working on producing tea picking machinery. The problem with mechanical harvesting is that it picks everything, and another day has to be used to pick out the trash, sticks and weeds. The university staff has come to my farm several times to try out and demonstrate some of their new contraptions. One was a handheld dust pan picker with a battery powered saw tooth edge. It was fun to use on perfectly uniform stands of new growth but was burdensome and useless with any irregularities in the growth. They also demonstrated a two-man machine that it took workers to pull it along the top of the row, which again had to be perfectly flat. This was major work just controlling the weight of the machine. They took lots of video of my hands picking and several videoed interviews as to why I did not want a mechanical harvester.

The Great Mississippi Tea Company now uses a large, heavy machine mounted on a steel frame with four tires. It takes two workers to push it along the perfectly box shaped rows. I said the workers looked like Egyptian slaves pushing blocks for the pyramids.

There is no question that handpicked tea is higher quality than machine picked tea.

Report on Side Growth in Tea Plants

(After putting information on my web site about how I was getting my tea rows to spread out, I received several requests about it, so I produced this report which I sent to tea producers all over the world.)

In the fall of 2020, I decided I needed to let my flat top tea bushes get wider, to produce a wider production table on the flat top rows. This was getting to be the style of tea rows that you would see in pictures of tea bushes most any place in the world. Particularly of interest were my good friends at The Great Mississippi Tea Company. His rows were low, flat and much wider than mine and my rows were 15-20 years older. So just getting older did not train the bushes to be wider.

Up until then, I had our yard service lady, prune the tea for me. She needed the wintertime work and she did not charge too much to do it. She would trim them hard in a rounded top shape and cut the sides like a Jim Nabor's haircut. So I instructed her not to prune off the side limbs that I called "phalanges." She really couldn't control her pruning instincts so that fall my tea again looked like well pruned box wood hedges.

The next fall, 2021, I decided I would do the pruning. Also, another incentive was our yard service lady charged me more than double than what she had charged before. So I did the pruning and left the thin, gangly side limbs on all the bushes.

Leaving the side growth when pruning.

That year I put a lot of thought into how would I encourage lateral growth. I even tried a long bamboo pole to mash down the side limbs to encourage them to spread out. That didn't work. Then after the next growing season began, spring 2022, I noticed how some side limbs would bush right on out with their auxiliary buds sprouting out filling up the space between the limb and the rest of the bush. That got my attention. How could I encourage auxiliary buds to grow?

With a lot of different plants, if you prune off the top, it encourages the sides to grow. Pine will not do that and if you top elderberry bush, growth will take place on another part of the plant. I even had my brother Dr. Jim Barrett, famed horticulturist, come and look at what I was doing and consulted with him several times about this.

In the spring of 2022, I laid out a demonstration to watch side growth, especially where auxiliary growth took place. Involving 22, fifty-foot rows, which all had lots of phalange growth, I would randomly not pluck, harvest the top buds of about half of the rows and the other rows I would pinch the terminal bud and a couple leaves. This was costing me lots of tea production by leaving so much good tea in the field getting to old to use, on both the unplucked and plucked rows. Before I would pick anything on top and sides, but I wanted to see what would make the side buds grow. I maintained this picking pattern all the summer of 2022 and ended up producing 270 pounds of tea that year.

By the end of the 2022 growing season (November) and also in the spring of 2023, it was evident what my results would be. On the rows where I had pinched the terminal buds off the phalange limbs half of the limbs had good auxiliary growth, half did not. On the rows where I did not harvest any growth off the phalange limbs, half had good auxiliary growth and half did not. This was so 50-50 it was conclusive that it really made no difference. But what I did see was the rows all got wider, some twice as wide. This was impressive and just what I wanted to see. So my conclusion of this demonstration on how to get your tea rows to get wider picking tables…just don't prune away your side limbs.

A wider row produces a wider production table and a lot more tea!

Note towards the end of 2023: Now that my tea rows are wider I find it makes them harder to pick the tea! The low flat rows are ok, as I can wiggle into the limbs to reach across. I find I can barely reach the new growth on the larger cupcake shaped rows and am actually leaving lots of it out in the field. I think we are going to do some hard pruning this winter.

Making Black Tea

1. Pick leaves
2. Wither
3. Roll
4. Oxidize
5. Dry
6. Toast

When we start to make black tea we have to plan ahead and let the tea grow a bit taller and older than we do when we make green tea. We are after the natural sugars and starches with green tea but we want the stronger, bolder flavors with the black tea.

Usually I pick the flat rows (Pancake) for green and the round top rows (Cupcake) for the black tea. I keep up with that for most of the picking season, plucking at the rate it is growing. Later in the year as the growth over all the rows is outgrowing my ability to keep it picked, I tend to skip around picking the age of leaf I want. When the growth gets too old, I pick around it and prune it off in the winter time. Also, the large cupcake rows (some call them Champagne glass) get too big to reach over and are hard to pick. They are also the worst for hiding wasp nests. With the flat rows you are able to wiggle inside and easily pick down on them. I only have seven of the big round rows left and we are slowly cutting them down to flat top rows.

When I pick the leaves for black tea I will bring them inside and spread out on an old bed sheet on our living room floor. Lottie will pick out the weeds, bugs and most of the tea twigs while I am picking. Then we start the withering step.

Withering is important. This is where the leaf moisture is reduced and the green leaf releases a bound up enzyme called phenyl esterase. I have observed several withering techniques over the years.

When I first went to the Chinese tea factories in 1984, obvious features were long, narrow lightly constructed buildings shaped like commercial chicken houses and mostly open on one side, covered in canvas drop cloth or plastic strips. These were described as "withering sheds." They had heating units on the ends with large fans blowing heated air into the building, like you would see in a large greenhouse. The building had rows of shelf/racks only a foot vertically apart and extended the length of the building. The racks would hold the five feet round baskets with tea leaves spread thinly, about two inches deep. Tour guides carefully explained how the withering was done in warm air for an appropriate amount of time.

At one site the workers spread out fresh tea leaves in the sun on large tarpaulins. I remember the guide with big teeth saying with a strong Chinese accent "Solar Withering." As I did not need to build a building, I utilized this technique and spread my tea out in the sun after picking for several years. This gave mixed results and sometimes twenty minutes would be enough and sometimes the leaf would dry too much and the oxidizing process wasn't always good.

Even the Sri Lankan Tea Institute recommended withering in heat. It was unquestionably the accepted technique until modern science taught us a better way.

A tea making friend of mine, who is a biochemist at heart, suggested a cool withering. This went against all the withering techniques I had observed, and I was slow to take to it. He said the warm withering started the oxidation process too early before the esterase was in place and oxidize and damage some of the flavors we were trying to hold up. Also, the esterase action reached its peak at 3 hours and that damage to flavors would take place after that. It made sense chemically, so I tried it and made the finest black tea I had ever produced.

I now wither in the 72-degree air conditioning for eleven to thirteen hours – overnight. During the night I will fluff the tea by throwing it in the air a couple times during the night (I can do this sleepwalking). I'll watch that the sheet it is on does not collect too much moisture, or I will change it. And, I always first put my hand in the heap, making sure the anaerobic bacteria is not being active, because undisturbed, they will heat up and start eating sugars and starches.

In the morning the sheets will be cool from evaporation and the little tea twigs will bend easily and there will be small splotches of brownish oxidation trying to get started. The leaves smell fresh and are cool to the touch.

Cool withering in AC around 72 degrees works well for us.

I then take the tea to the rolling machine. We roll tea differently for green and black and according to the age of the leaf. Early in the year, if you roll the fresh new growth very hard at all it will make the tea cloudy. Late in the year you will roll both green and black really hard to curl the older leaves. I generally roll black tea harder because it encourages the oxidation process. I back off on green tea so it will be a little bit more leafy.

So, with black tea, I will fill the drum or hopper up and just bring the lid down to the level of the leaves. I will roll this softly for 10-15 minutes. If I have more to roll I will take this soft roll out of the drum refill with fresh leaves and soft roll this batch too. If needed, I will do this soft roll a third time, depending how large of a pile I am processing.

After the whole pile is soft rolled, I start feeding it back into the drum, pressing it down and then screw the lid down hard and roll for five to ten minutes. This makes a big difference in the leaf appearance. It is crushed up more and the tea leaf juices are making it wet and shiny. This hard roll also shrinks the volume so I will open the drum up, break apart the knots, press down and fill on top of this more softly rolled leaf. I will put the lid down firmly and roll another five to ten minutes. And maybe again a third time until I have it all in the roller and finally rolling it all again with the lid firmly screwed down. This produces a leafy mass that looks like greenish colored Cole Slaw. You cannot do this by hand, no matter how hard you work at it.

Rolling black tea firmly smashes the leaves up to looking like wet Coleslaw. This helps with the oxidation process.

Then it is time for the oxidation/anaerobic bacteria step which produces the black tea flavor. This is where the phenyl esterase produced during the withering step will start to oxidize the polyphneniols, natural plant sugars and starches which produce the all-too-familiar flavor of black tea. It also is the time the anaerobic bacteria, there are about 50 in our environment and some 15 come into play with tea, produces the sharpness, acidic bite of black tea. These two chemical reactions are essential to the black tea making process.

Rolled tea leaves during the oxidation step, starts green in color and turns brown.

We will remove the beat up tea leaves from the roller and put in a plastic tub. I have several size tubs to use depending on the size of the batch. For a long while I used commercial ice boxes as you would see on a shrimp boat. As these were getting worn they would leave small balls of Styrofoam in the tea which looked nasty. Some people use ice boxes with heaters and put the tea leaves in glass casserole dishes. This does not compact the tea for good bacteria growth and they also will fluff or roll the leaves (?) during the oxidation step which will inhibit the bacterial action.

Another tea maker rolled the leaves in a towel and tied it up hard to facilitate the process. In China they called these "dragon eggs." I did that for a while placing the ball on the hot water heater. This worked well but my volume of picked leaves had me trying the ice box and tub method.

I will mash down hard as I pack the tea in layer on layer, smashed down as firmly as I can pack it. Then I would wash a towel in hot water and mash it down on the leaves, tucking it in around the edges and mashing again down hard to squeeze out the air. Then I put a thermometer under the towel. Because the roller condenses the leaves, you can get about three bushels mashed down tightly in a five-quart bucket (shrinking as when cooking turnip greens). I now like working with smaller batches to help control the heat.

The formula for oxidizing black tea is three hours at 95 degrees. People pay money for those numbers. When I began making black tea, I was copying what I saw in China and it appeared they let it stay in the oxidizing stage for a long while, so I did this. I was never satisfied with the results. I gave a tea distributor a cup of my black tea (around 2000-2001) and right off he said it was "flat and stale." By experimentation and accident one day when I was in a hurry, I stumbled across the fact that if I let it oxidize for only a few hours the tea was sweeter and much better. Years later I saw a graph on "Lets Grow Tea" Facebook page that showed the flavors in tea were being oxidized and suffered after three hours.

I used several heat sources to get it warm. First, I would use the top of the hot water heater. Then for a long while I would bury a hot water bottle (or two) down in the leaves and with the natural propensity to warm itself, this worked well. Then I started placing the tub in the sun and watched the thermometer, sliding it into the shade when it got too hot and then sliding it in the sun when it got cool. I do this today to keep it in the 95 degree range for three hours. I also have a small single bed size electric blanket dedicated to heating the tub up during the still cool spring months. On the "8" setting, the tub rolled up in the blanket will hold steady at 95 degrees.

The black tea leaves after oxidation.

After the oxidation step, I take it to the dry pen. I spread the hot, wet, smelly tea out on old bed sheets with china dishes on the corners. I will break up the knots with my fingers and spread as thinly as possible. I will rake it around with a dedicated rake and within hours it will start to dry out. I likely will let it spend the next afternoon in the dry pen where it will become crunchy dry. Another day topped off in the air conditioning and the tea will be ready for the oven.

Black tea drying in the sun.

We then toast the black tea. This is not drying the tea which produces a metallic taste to the brew. We do not toast green tea, which would damage the sugars and starches, but we do toast black tea, which fractionates the flavonoides. The toasting clearly makes the tea more zesty and favorable. Folks who dry their tea in an oven miss the whole point of this.

I did use old ovens ($50 each) from an appliance junk yard for years, keeping the old ovens in the barn and bringing one up to the back porch when needed. It started being too much trouble to clean them up for use, so my wife and I now toast tea in our kitchen, making sure we don't make a mess.

Lottie and Jeanine Normand toast tea while museum docents in the background rub it through a screen (this is Oolong tea.)

We have large commercial cookie sheets which are the size of the oven which we spread the tea on thickly, maybe two inches deep, for toasting. The density will vary the thickness and times. We will toast around three minutes with the oven setting between 225 – 300 degrees. It changes in texture and color quickly so you watch it very closely, never burning it.

After the tea is completely cooled we will pour the tea into large five gallon Mylar bags. I pound the tea in tightly using a gloved fist which does break up the leaves. We then add an oxygen eater packet then iron closed to seal.

These weigh 12 -14 pounds. I keep these in my air-conditioned store room in the barn at a consistent temperature.

Mylar bags of tea in the storeroom.

Making Green Tea

1. Pick
2. Cool leaves
3. Pan fry
4. Roll
5. Sun dry

Lots of tea producers in this country cannot even define what green tea is. A Facebook page article, not long ago, contained long discussions on what makes green tea green. Seems most American tea makers don't like green tea, work to change its character or don't make it or understand it at all.

I had four distributors to visit who worked for the same company and flew all over the world buying tea for their company and their tea blenders. They knew much more about the world of tea than I ever will. I asked them to define green tea, and they got in a spitting match among themselves trying to agree on what green tea is.

We begin with planning ahead for pruning and picking the leaves for the green tea. We prune the rows low and flat in the spring and pick the new growth about every two weeks to get just the top of the growing bush. It is a bit more than the "bud and two leaves" we hear in commercial ads. This young growth has the sugars and starches we think is essential to the green tea flavor.

I will pick the leaves while my wife is inside processing it. It is just that sensitive. As soon as I pluck a leaf, I am making one type of tea or another. When I get about half a bag I bring it inside and we spread it out on the floor on old bed sheets and toss it into the air several times to fluff it up. This stops the anaerobic bacteria from attacking the natural sugars and starches.

Have you ever bought a bushel of butterbeans and did not shell them quickly? Put your hand down in the basket and the beans are getting hot and moist. That is the anaerobic bacteria action. There are about 50 varieties of these bacteria in our environment, part of the decomposition process. Some 15 varieties of these bacteria come to play in the tea making process. These are what gives black tea its bite, acidity and sets up its flavor. We want to stop this when making green tea.

I first read about the bacterial action in tea making in the publication "Twenty-five Years of Research at the Sri Lankan Tea Institute." It is surprising, almost startling that most American tea producers do not acknowledge and ignore this important aspect of tea making. One of the tea consultant agencies that sell their technical services to new tea makers completely avoids the subject. They now have trained tea makers all over the US. You can do things to facilitate this bacterial action and there are techniques you can do to hinder it. Again, we want to stop this anaerobic bacteria action for green tea. Simply spreading it out in the AC will do it. If you are a large commercial factory you can't, don't or won't do this.

After we have cooled the tea, then we heat treat it, which sounds like an oxymoron. This is to stop a chemical reaction. My wife will heat up a wok or large electric fry pan to heat the tea leaves. This is not cooking or drying the leaves but just heat treating. Around 170 degrees will do the job. For a few years I would steam the green tea to stop oxidation. I had made a hardware cloth flat basket that opened like a clam shell with two halves. I was mimicking a steam machine I saw in China. I made this the size of a large stainless army cook pot and would steam one side then flip it over to steam the other. This involved lots of work loading and unloading the basket. Also, the results were uneven with leaves in the middle not getting hot enough. Somewhere along the line someone or something told me pan frying produced better tea so I changed to pan frying.

Lottie "heat treating" the green tea.

This heating or steaming the green tea leaves stops an enzyme chemical known as phenyl esterase. This natural occurring enzyme is in all green plants and is released when the stem is broken. Its purpose is to oxidize sugars and starches (remember we want to keep these in green tea) into linear hydrocarbons, esters, ethers, aldehydes and alcohols, to stimulate the growing of roots. This is very similar to the aromatic chemicals produced when toasting a piece of bread. The volatile hydrocarbons are stimulating. This builds the flavor of black tea (oxidizing polyphenols) but we stop this for green tea, the heat treatment stops this oxidation reduction reaction. Then we take the pan fried leaves out to the tea roller.

Green tea after being pan fried, ready for the tea roller

After the rolling, we take our rolled leaves out to the dry pen. The dry pen has solid plastic walls to stop the wind from blowing across the surface. This produces an almost microwave effect on the ground surface where we spread out the tea in the sun. This is how they dry seaweed, kelp, doltz, apples and other fruit. I saw these small dry pens built of masonry walls in China. That's where I got the idea to dry tea in the sun. We spread the tea out on old bed sheets with old dinner plates on the corners. We use the plates to hold the sheet in place from raking and not the wind. We use plates because you can see readily that they are clean. You can't look at a rock or brick and tell if it's clean and anything on the weight will go into the tea.

When we bring green tea directly from the roller and place in the sun, the tops of the gobs or wet leaves will "sun tan." This is a little bit of oxidation which I think gives green tea a bit of boldness. As this tea has been heat treated, this is not the same oxidation as if it were black tea, it is only slight. If you wait a day and then put out in the sun (as if when it is raining) it will not sun tan. Sometimes I will do that on purpose to give a blending of tea batches, some sun tanned and some not. I will then rake it around with a dedicated yard rake that I would not use to rake leaves in the yard. I also, at this point, break up the knots in the tea produced by the roller. I have no explanation why the green tea gets more knots than the black tea.

I would dry the leaves out in the sun and keep an eye on them. I will put the leaves out when the morning dew has gone and be sure to pick up before the evening shadow fell across the tea. If that happened, the drying sheet would quickly absorb moisture from the grass and quickly moisten the tea. This has a downer effect on the flavor, same as if I forgot about it and let it get wet in the rain.

A batch of green and black in the dry pen.

Every day when I brought the tea in from the dry pen I would spread the sheet out on the floor in the air conditioning but leave the tea in a ball in the middle. After the moisture had dried out, out of the sheet, I would then spread the tea out. This is artisan.

After a few days in cool, dry air conditioning, the green tea is ready to store. For years I used old coffee cans, the metal kind with snap on plastic lids. When they quit making those I started using the all plastic two pound cans and managed a large pile of these from family and friends. I would pack the tea in the cans with a wine bottle.

When the green tea is crunchy dry to the touch, which sometimes only takes a good drying afternoon in the dry pen, then I will bring it in the house and spread it out in the air conditioning. A couple days of this will get the tea leaf moisture down to about five percent. This is when it is stable and stores well – indefinitely.

We do not toast green tea. This would damage the sugars and starches and the whole point of making tea green is to preserve sugars, starches and essential oils. Most producers dry their tea in an oven. This gives a distinctive metallic taste to green and black teas. I did several side-by-side taste tests, serving tourists a "cup A" and a "cup B," and asked them which was better. They all liked the sun-dried tea better. I never dried tea in the oven again after that.

Most producers also wither their green tea. They do this because someone instructed them to do so or they don't really like green tea. When green tea is first withered it produces the phenyl esterase which immediately starts to oxidize the grassy sweetness in the leaf (I minored in college chemistry). The major component of this grassiness is several flavonoids, especially the gallocatchins. These are polyphenols that give green tea its bitter-grassiness and astringency. Some people, like me, like that, but most producers don't so they wither their green to oxidize the polyphenols into the different flavins. So green tea will be grassy or earthy, depending how it is made.

Green tea that has been mildly withered is actually known in China as pouching. If you wither the leaf more aggressively it will become Oolong, which is about half green and half black tea.

This Oolong style of tea was what we sold many thousands of dollars of while I worked at the Fairhope Museum. We would wither it at the same rate as producing black tea but then we would pan fry to stop the oxidation. You had to catch the processes just right which was tricky at times. We also would toast (and screen) the Oolong. We did not toast our green teas.

This is me screening the Oolong tea.
We don't do this much anymore.

Yellow Tea

In 2020, I was approached by a couple of men in army uniforms, in a black limousine, with rows of polished brass buttons here in my drive way. They requested I make an appointment to meet with a Chinese agent from the Trump White House in Washington, DC.

After several confirming telephone calls, the Chinese gentleman arrived accompanied by two CIA agents. It was all rather intimidating. He announced himself (I won't use his name as he may be back in operation) and presented me with very nice White House gifts, and had me promise not to discuss what he did in Washington.

It seems he was from a tea making family in China that made green tea. He was not concerned with black tea, what the Chinese call "Red" tea. He was highly irritated that Trump had lost the election as he was considering bringing his family to Washington, but was now headed home.

He held Americans in very low esteem. He thought we were unmotivated, uneducated, vain, greedy, self-serving, disrespectful and vulgar. All this may be true but he certainly did not know how to not say it!

We disagreed on everything. He stayed in a local hotel for three days and spent most of his time with me discussing tea. The two CIA men watched him the entire time. We once went to a restaurant and they came in and sat at a table near us. It was rather creepy.

The Chinese tea maker would hold up a green leaf, snap it in two and say "You do that, nobody want!" Chinese prefer whole leaf tea. Americans don't know what to do with whole leaf tea. I make some every year but eventually crush it up into teabags. When I was packing tea in cans with an empty wine bottle it made him roll around in anguish.

I don't think he heard anything I ever said, but he did show me how to make yellow tea the way his family made it back in China. My yellow tea is by far my best tea.

There are two ways to make yellow tea. One is taking the fresh green leaves and putting them in a plastic bag and sucking out the excess air, seal, and then putting them in a cool place for several days. This allows the anaerobic bacteria to digest sugars and starches, part of the decomposition process. Some producers will occasionally open the bag, shake out the accumulating noxious gases then reseal, and some don't. The deteriorating leaves will turn bone yellow and the taste is strongly sharp. I made this style of tea once, tried it, and then threw it into the compost heap.

 The Chinese gentleman taught me how to make yellow tea in the style his family makes it. This is tea made without heat or sun. His pickers would pick the very first flushing buds with one or two tiny leaves. They would place on racks, without the buds touching and air would be circulated around the trays. Within the first two days, the tiny tea buds become a bit sticky and can be rolled into little balls.

I simulate this process by picking the first spring time emerging swollen buds, which involves a lot of picking to produce a small bowl of green buds and tiny tea leaves. I spread them on a cookie sheet, without them touching, and let dry in an air-conditioned room, actually sliding them under the bed and forgetting about them.

This tea is very good with a nice fruity aroma, but we can only make about five pounds of it a year at springtime, not enough to package and sell. This is the tea I brew when a customer – maybe twice a year – requests gongfu style or whole leaf tea.

Our yellow tea looks much like other producer's white tea. This is our best tea!

Drying Tea

"...and, then you just dry it out,"

I hear this over and over. Folks think "all-you-gotta-do" is pick the leaves and dry them out. This will make dirty water that smells a bit like tea but it is not amber colored and has a completely flat taste. Yes, tea is a dry commodity but you have to go through several important steps before the tea is dried.

I saw several drying techniques when I went to China in 1984. I saw large ovens for toasting and/or drying but mostly they were using the sun. Even the large factories had areas where the tea was spread out in the sun. I saw the dry pens like I use, but they were brick or cement walls and there were rows of them. I saw large screened in rooms with a screened roof that let in the sun but not the rain. In several areas I saw the Chinese workers stop cars (very few then) and wagons and sweep the black top pavement and dry tea on the hot road bed. Traffic was diverted around the drying tea on the side where the wind would blow the dust away from the tea.

When I first started making tea, I would put it on a cookie sheet and toast dry in the kitchen oven. This got to be a bottle neck and I started drying tea in the sun. I quickly noticed the difference between the two with the oven dried tea having a slight metallic taste, because of the longer time in the oven. So, I have used the sun to dry my tea all these years. When I get samples of other maker's teas, I can always taste their drying oven. We will toast our black tea in the oven for 2-3 minutes, but this does not produce the oven taste.

I consider drying in the sun to be "artisan." It does take planning ahead and watching the weather. You have to wait until the dew is gone in the morning and you have to pick it up before the afternoon shadows cover it. The shade will cause the moisture in the grass to quickly dampen the drying sheet which, same as a sudden shower, will produce a musty smell and taste.

For years I used my old Civil War reenacting wool blankets to dry tea. I now use old bed sheets which give up the moisture much faster. I look like Santa Clause carrying a large bag on my back as I haul tea around. We will get several sheets of tea on our floors while it is being processed. Most tea producers use large flat baskets for this but you cannot carry them loaded with tea through a door way.

The dry pen works well. You can bring out a batch of freshly rolled or freshly oxidized tea, which will look like wet coleslaw, and spread it out thinly on an old sheet. By four o'clock in the afternoon it will be crunchy dry. Sometimes it gets another turn out in the sun but usually one day will do it. Then another day or two in air conditioning and this batch will be ready to bag up or toast. We use five gallon size Mylar bags which we put in an air eater and iron closed. This will be about ten pounds

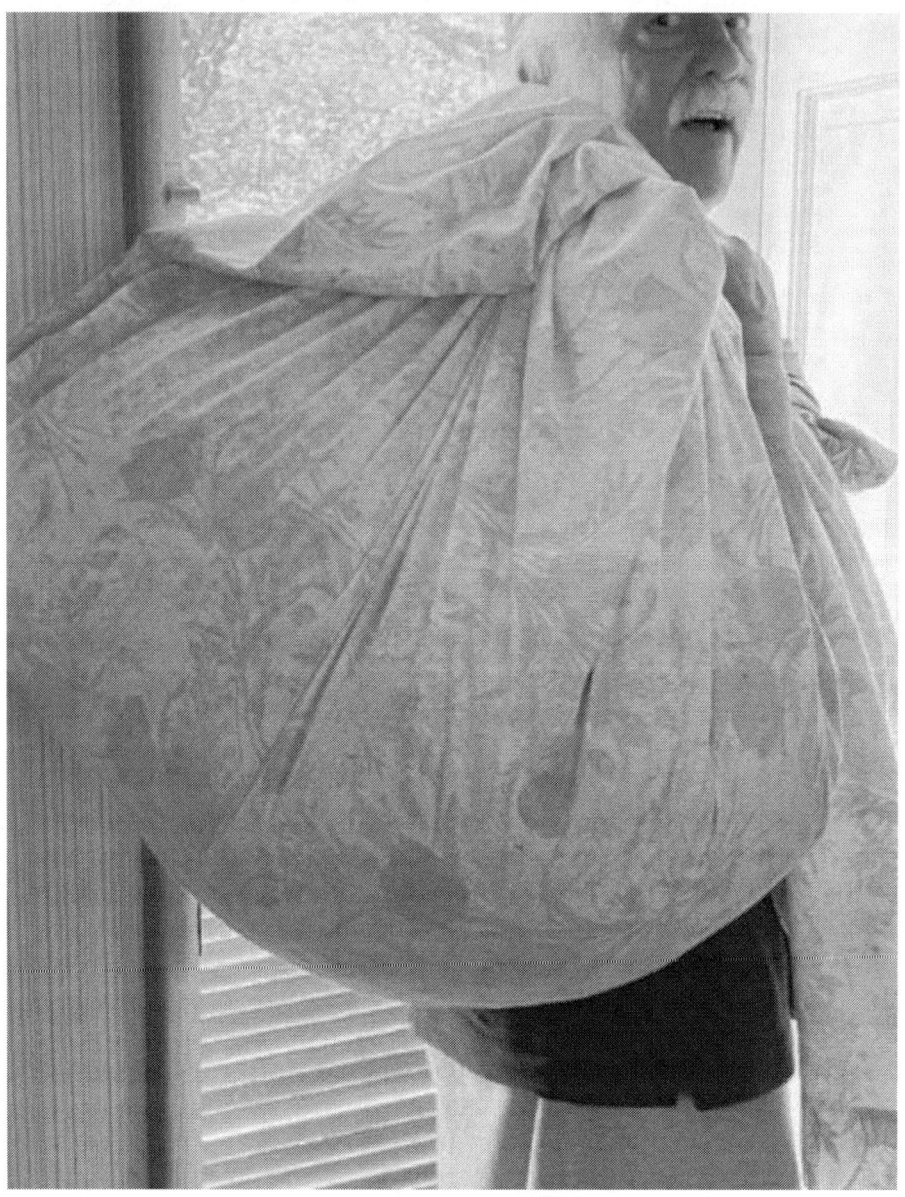

of green or fifteen pounds of black. *We use lots of old bed sheets.*

Blending

The brew from tea is like the wine from grapes, no two batches are the same. The batches made the same way closely resemble much of the time but then nature throws out a taste or feature a flavor that is unexpected. These unexpected flavors are quite different and surprisingly new. You can seldom recreate a new flavor you may find to the liking. I notice this most all the time with Lottie and I sitting on the couch, sipping a fresh brewed pot and saying "now, what did we do to this?" or "this is different than yesterday's, what tea is this?" We often disagree but admit we wore out our tea tasters long ago. I will actually serve our guests a cup "A" and a cup "B' and ask which is better.

Tea is blended in all sorts of ways. I have had professional tea blenders, who make a living knowing and blending different teas, try my tea. They said most all the tea you can purchase commercially has been blended. The strong flavor of English, Irish and Scottish breakfast teas is a product of blending.

These professional blenders are focused on the different attributes of different teas, and then try match up teas with flavors favoring the opposite ends of the spectrum. Such as a dry, robust tea with another with strong flowery attributes, opposite flavors that will complement each other. I was describing this to a guest one day who happened to be a wine blender for the Gallo Wine Company, and she said, "Oh, Honey, you have no idea!!"

We have never blended anything with our teas. Everyone else certainly does. I've never tried peppermint, jasmine, tree bark, chrysanthemum or pepper spice or anything else in tea. The only blending I have ever done is blending a bit of green tea in my black tea and a little black tea in my green tea. Sometimes I will use just a little and sometimes as much as a 50/50 blend. Green tea is made saving sugars, starches and essential oils – strong on the flowery end. Black tea sacrifices some of its sugars and starches during oxidation to produce the robustness and bite of black tea. Together they complement each other.

I actually discovered that by accident. When we did bus tours for as many as fifty-five people at a time, we would make a large pitcher of green, Oolong and black tea for them to try some of each. After the party we would pour what was left together for us to drink. "Wow, that's the best tea!" We now have had tea blending professionals to come here to visit and tell us that is no accident.

We also have found that green tea is more flowery and brews more cloudy in the spring. Through the year it loses more of the flower and becomes more grassy and a tad bit more bitter. So when I am making teabags, I will add a scoop of April's green tea, then a scoop of August green tea, then a pinch of black tea. That makes a good brew!

Another technique they tell me about is when they make the "Cut, Crush, Curl" tea, which is now called "Cut, Tear, Curl" style tea which produces a processed food. An employee of Red Diamond from New Orleans came to visit and told me what they did. They will crush the tea when picked, put in large steel tank with water and heated to the 104 degrees which oxidizes the tea quickly. They will run some of the liquid out, place in a spectrophotometer and look at the six basic flavonoides. They will add to each a chemical flavor derivative to bring it up to their standard. Add food color so it will always look the same then dry the liquid back down onto the tea particles.

So, you can take their teabag and make a cup of tea in a minute and a half but you can only do it once. You have washed off the residue stuck to the particle and not leaching the flavor from the leaves – a processed food. That is why "sun tea" works. You could put my tea out in the sun for two weeks and it would not make tea. Real tea needs heat to remove the complex flavors from the leaf particles.

Another practice we have been using is making our tea, loose leaf, very strong. Then we mix it with hot water down to taste. The strength of the different batches can vary widely so we make it strong then water down to the right strength. I tell this over and over to my customers, but they look at me with a blank look because they are not getting what I am saying.

Loose Leaf Tea

Most cultures all over the world drink their tea made with loose leaf tea. In the early 1900's Mr. Lipton started putting samples in cotton, muslin bags. People started putting those right in the hot water, which was not his intention. Lipton did not invent tea bags but he sure made them popular. So, traditionally, England, Canada and America use tea bags, use black tea and call it "tea." The rest of the world calls it cha, uses green tea and uses loose leaf. Although tea is relatively inexpensive, you now can buy any tea in any form from any place in the world, loose leaf or in teabags.

Most households now have a "tea ball" for using loose leaf. It is an aluminum ball with little holes all over. It is hinged so it can be opened up and loose tea placed inside. This keeps the tea from freely floating around in the water. I call them "anal." People are proud to say it was their grandmother's. Do you think she ever used it "no." Have you ever used it, "no."

I make all of our tea with loose leaf tea. The only time I would use a teabag is when I'm running a test of some sort after changing bag size or amount of contents. With

We sell about half of our tea, green and black in half pound loose leaf bags.

a side-by-side comparison, loose leaf makes a better cup

of tea.

I use a two-quart kettle which I fill to ¾ full. I heat until small bubbles start to form then I turn off the heat. I will pour ¼ cup loose tea in the water and let it float around and steep. After five minutes I will put a "reusable coffee filter" on the mouth of a pitcher. These are plastic funnel shaped cups made with a fine mesh screen designed for a coffee pot to replace paper filters. One will last many years. I pour the tea with floating particles through the filter which catches any that comes out of the kettle. I then take the kettle and screen outside, wash it out in our outside sink and throw the residue over the fence. This works very well.

We sell half loose leaf and half in teabags. We sell 50% black and 50% green tea, so we make half green and half black during the summer and it works out fine when we are down to very little, near selling out in the spring. We also make a few pounds of whole leaf tea for gongfu drinkers, but have little-no demand for it. Towards the springtime I will crush this up and sell it in teabags.

Brewing

This is one subject that I would seldom take up with tea tour guests. Folks have their own way of brewing tea and they are going to buy my tea and brew it their own way. If they ask me, then we can discuss it but otherwise I sell the tea and off they go.

I tend to brew my tea pretty much the way my mamma made hers. She would put two Lipton "family size" tea bags in a ceramic tea pot, the same one she used when I was small and still used until she died in 2009. She would boil water in a copper kettle then cover the teabags with hot water. This would make a hot, thick concentrated tea. After the appropriate steeping time she would pour this concentrate into our stainless steel, Army surplus pitcher. She would add two large scoops of sugar and stir. Then she would add cold water to bring it to the desired strength and cool it a bit so as not to melt the ice in the glass. She would be real fussy if a "helper" interfered in her tea making process. I never saw her sip hot tea from a cup. It was all about sweet, ice tea.

My style of brewing starts with a two-quart kettle. I fill it three quarters full with water and heat on stove to almost a boil. If I forget it and it boils vigorously for a while I will take the kettle out and pour the water on an ant bed and start over. If water boils for a few minutes it will produce a flat tasting tea because the air has been boiled out of the water.

When the water is almost boiling I will pour in quarter cup of loose life green or black. I almost always put a pinch of green in my black tea or a pinch of black in my green tea at this point. I let it steep for five minutes (Lottie steeps for four minutes) then pour it through a nylon screen over the pitcher. We use the nylon baskets designed for use in coffee pots without a paper filter. We have used these for years as they fit well on the top of the pitcher.

I also make this very strong with lots of tea and long steeping times. I then put hot water in the pitcher and water it down to taste. All of our teas have different strengths and we go from one batch to

My mother's tea pot.

another and this watering down technique helps us get it right.

I had a Chinese visitor, Widgeon He, an agent in the first Trump White House to scorn me for breaking up the leaves and making a large kettle of tea as "Oh, so American!" He would hold up a leaf, tear in half and say "you do this, nobody wants." His family was green tea makers in China and he seemed to care little for black tea. He referred to my tea as "raw, cut leaf." This is a derogatory term I have heard also on Utube meaning little twigs and bits of non-compliant leaves (leaf that did not curl) were present and the leaves were broken into small pieces. He cringed when I packed tea in a can with a wine bottle.

Chinese use whole leaf tea and brew in a gongfu (or kungfu) style. This involves using a small teapot known as a gaiwan and small cups without handles known as bezee or chawau. Whole leaf tea is used with multiple infusions, with the first usually discarded. The different brews are described in an array of flavor descriptions. I make some whole leaf tea every year but have no market for it. Maybe I will get one or two requests for whole leaf in a year. At the end of every year I will grind it up and put in teabags just to move it on. I'll have to admit that the taste of gongfu style tea is better.

This six inch gaiwan was left at my house by my Chinese friend Shen Hua who lived with us for a year.

This is the modern version of a gaiwan.

Pests

I've not had to deal with any insect pests or diseases in the tea farm. I have never sprayed the tea for any bugs or diseases. Yes, we have some down-right irritable insects but none that are eating and killing the tea bushes, they mostly eat me while plucking.

For several years I had a professor from the University of Florida to come here several times a year to study and take many samples of the tea mites that live on tea. Because I have never used insecticides, he loved my mite community and had lots to say about it. Dr. Childers said my good mites were eating the bad mites, a sign of a healthy ecosystem. He went to every American tea grower he could find to study the tea mites.

In foreign countries they certainly have other insect problems. A stem borer in other countries around the world is the Shot Hole Borer. We have stem borers here that will kill your squash, gourds and pumpkins but they do not attack tea. Shot Hole Borers are referred to in scientific publications from around the world. It bores into the new growth and sucks the life out of the young stem which then withers away. Pesticide use is prevalent in most countries known for tea production.

A tea distributor in New Hampshire that I was selling tea to in the 1990's was also a biochemist. He would do chemical residue analysis on most all the tea he bought from all over the planet. He said he would find chlorinated hydrocarbon residues in most all his samples, some at an alarming high rate. These type insecticides are not permitted in the US but are in other countries around the world to fight the Shot Hole Borer.

In our shade-grown rows, the tops of the plants will be home to mealy bugs. These are not a problem and if I wanted to pick the area where they are and use the tea I can wash them off with a garden hose. Also in shaded areas, the very small seedlings could have a little scale under the leaf but doesn't stay present very long.

Fire ants are a constant hazard. They do not have to be in a big soft mound but can come at you by the thousands by surprise. They are really diabolically smart. They will come up both legs without a bite and then on a given signal, attack your legs all at once. That is why every time I pick tea or just pull weeds, I spray Off repellant on my socks and ankles up to my knee.

Spraying bug repellant on my legs also helps deter ticks. I always shower after being in the tea because of ticks. I have had Rocky Mountain Spotted Fever once and Southern Tick Associated Rash Syndrome three times. These come from deer ticks with the little round red bodies and black legs – we have those out in the tea but not much anywhere else.

I think my worse tea insect nemesis are Paper Wasps. A fifty-foot row can have as many as three nests, almost always on the south side of the row. I carry ribbons in my pocket to mark the nests. I have been dealing with these for years and have developed a routine which helps me not get stung. My wife calls me the "wasp whisperer" because when I first encounter a nest with a hundred wasps I calmly (even if stung) step to the side and continue picking. This lets them know I am not a threat like a cow browsing. If you run, screaming and waving your arms. it excites them to attack more and gives them a sence they ran away the threat. The next time I approach the ribbon I just skip over it in a non-threatening manner. After a few times they are not threatened by your picking and don't even fly around. I took me ten years to learn how to do this.

I do not fight weeds in my tea rows. I think tea likes the competition. If you let weeds grow in your garden they would kill your vegetables but tea is a climax forest plant and will easily out compete any weed. I think tea also likes the shade from weeds.

There are a few weeds that do present a problem. *Solidago*, Goldenrod and Horse Tail become a problem when they get six feet tall and flower. Picking around these tall flowers causes trash from the flower mass to get into your tea that you are plucking.

I used to pull these up getting a truck load of debris to remove but now I just pinch off the tops of the plants as they are growing. This makes them struggle to fork a little and have a small flower down low. Also after doing this for several years I have found the plants don't reproduce well and now I have little *Solidago* when I used to have it all over the tea field. A fellow tea producer that I had shown this technique, and others, went to Hawaii where it was published in a tea farm manual as the "Barrett Weed Control."

Just two years ago a strong growing, six foot weed suddenly appeared in my tea and all along our roadsides. I asked several of my botanist friends about this and no one could identify. My horticulturist brother, Dr. Jim Barrett could not identify but said "it had a mint leaf." After several searches a forestry professional told me about Tropical Bushmint. It was blown ashore on the Alabama, Mississippi and Florida beaches during the 1990 hurricanes, where we had a category 1 storm for eight years in a row. It is now only found a few miles from the coast but being very aggressive it will be coming to a fence row near you soon. It stands and looks like Ragweed (I call Ambrosia with my strong Southern accent) but grows thicker and dominates the plants around it.

A new invader, Tropical Bushmint.

Briers are a nuisance especially when you grab some unseen while picking. If there are lots, I will put on work gloves and tear the tops off where I pick. It would be more logical to pull them up by the roots at ground level but I don't often do that. I just torment them at the top. One thing I have found if you pick up a long sprout by the tip, there are no thorns there. You can lay it over upside down and it will die.

I have several grasses that grow in my tea field. The Vassey, Johnson, Broomsedge and Fescue grasses are not a problem. The Bahia grass has been a problem. It will grow up so high that it is up in the tea flush I am picking. It is so small you can't pick around it and very hard to pick out of the finished tea. I have abandoned several rows during picking because of the tall Bahia grass. It somehow seems not to be a problem anymore.

Cogon grass is my latest problem. I've gotten several patches of it in my tea rows. I am patient and pick the tea around the large, tall grass blades but it sure slows me down. We have had a real serge of it this year. I am going to have to contemplate some sort of herbicide. I would use it late in the year after harvest time then prune away the tea to remove any

Cogon Grass

residues before the next years flush.

We have two diseases present in our tea rows. Neither presents a problem and has never warranted any treatment. One is Tea Spot. This is where an algae will live in the stomas of the leaf and leave a small round circle dead spot. This is found on older leaves and not a problem. If you have a camellia japonica in your yard, look at the bottom older leaves and you will see tea spot.

Tea Spot

The other disease we have is Brown Blight. It will cause the tip of the leaves to die and turn brown. It has been a major problem for some tea growers but not here on the Fairhope farm. Most of the time I cannot even find any of it.

Fertilizer. I will go years without using any chemical fertilizers. These will stimulate the grass and weeds and show little effect on the tea. I went eleven years during one time period but usually go around five years before I use any on tea. I was a hippie in the 1970's when at Auburn University we were led to believe fertilizer was poisonous. I still think that way so I use very little anywhere here on our farm. If I do it gets the lightest sprinkle, a fraction of the usual farmers apply rates. When I cut an old row down to six inch stubs, I will fertilize these but I don't prune this way but every few years.

Young tea plants do not like fertilizers. They will develop yellow and brown "leopard spots" before dropping the leaves. Our good friends in Mississippi seem to use lots of fertilizers applied two or three times a year. I had one certified tea taster said not only could he taste it in a sample that had been fertilized but can open a container and smell it.

A soil test specifically for tea recommends a heavy application of potassium – Potash fertilizer. I put some on once but only once.

Tea Rolling Machine

All tea gets rolled. It facilitates the chemical reactions you are wanting and it concentrates the finished product. It is either done by hand or by a machine.

My first teas were hand rolled. I would put an amount in a hand towel and roll the softball size ball around on a table, even hitting it hard on the table. This really did little to roll the particles into nice fannings. For a while I would actually grind my tea up in a food processor so it would look more like rolled tea. I was then shown how to roll tea on a flat table with your hands out flat. A good varnished table worked well as the surface was a little sticky which helped to roll up the tea into little balls. A tea making friend does his tea rolling like this currently. I too will still use this technique when rolling yellow tea because it is just the bud and a tiny leaf and not near enough volume to

Rolling by hand, lots of work!

use the tea rolling machine which would actually just tear it up.

For several years I would roll my tea up in old tee shirts which produced a nicely finished tea, but a lot of work. I would spread the tea leaves thinly on the tea shirt then roll it up like a sushi roll. I would let it sit for a while then do it again. I would end this process by tying the cloth into a knot and let it sit overnight. I had a basket of these old tea shirts which were heavily stained. A tea shop owner offered to sell these in her tea shop with their unique look.

In 2019 I bought a used tea roller from a company in South Carolina, Table Rock Tea Company, Steve Lorch owner. I bought it for $1000. It was previously owned by a Yupon tea maker in Mississippi, who bought it then found out they didn't need it, so the machine was brand new when I bought it.

All of our tea now goes through this roller machine. It will do in 90 seconds what hand rolling would do in an hour. I could not make the

Tea rolling machine

amount of tea without this wonderful machine. It is well designed to do its job. It produces finely finished teas as compared with any commercial teas from all around the world.

It took lots of practice with this roller to get the texture we wanted. I was fortunate to see a Chinese Utube video on how to fill, how long and how hard to roll the different teas for the desired effect. The lid will press on the rolling leaves with any amount of pressure needed.

I have found out the hard way that if you roll softer, early season pickings too hard, it will make the brewed tea cloudy. Later in the year you need to roll the teas really hard, especially the black teas with the lid firmly down.

So, generally, I will roll green tea softly without much pressure for 10-12 minutes. I will open the hopper and fluff up the knots, then roll it sorta-hard for only five minutes.

With black teas I will roll it softly for 15 minutes then again roll firmly with the lid down hard for ten minutes. All these times and pressures are a learned skill that is tweaked for different batches and different ages to the leaves.

The tea will tie itself up in knots inside the hopper which have to be loosened by hand. If not, they will make a hard ball in the finished teas which is hard to dry. I had a reporter here one day and I referred to the knots as "snot wads." She said "Mr. Barrett, you can reassure us that no actual snot was used in making this tea?"

I will roll different teas in different ways.

Tea Bagger Machine

In 2020 we bought a tea bagging machine. It is made by the Huadacn Pharmaceutical Machinery Company in Ruian City, China. It came through Table Rock Tea Company, Columbia, SC, which is a vendor for many types of machines for what they call "mini-industries." They are also very helpful with technical information and trouble shooting.

The tea bagging machine makes pleasant sounds like a slot machine and does a really good job making teabags. You can fill the supply bin with tea and go off and leave it and it will turn off when done. We make around 500 teabags at a time and have made many thousands of them. They do not have strings or tags and are not put in paper sleeves.

We will grind the tea finer in a Black & Decker food processor which allows the tea to defuse better when

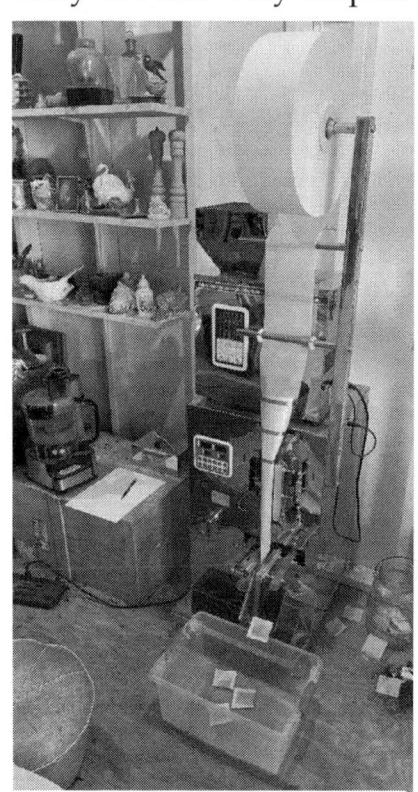

We make about 17,500 teabags a year.

brewing. "Orange Peko" which most folks think is a variety, actually refers to the size particles, fannings, which are just right for filter paper teabags. It is also used to describe different leaves on the tea bush and is also used as a commercial company name. We are currently putting two grams of tea per bag which measures two and a half inches square.

My first teabag! December 15, 2020.

We find that we sell half of our tea in teabags and half in loose leaf. It seems younger people like the teabags and more experienced folks like the loose leaf. In doing test runs for different teas and different times, it is clear that loose leaf makes a better brew.

Our tea bags are sealed with a strong crimping action heated to 350 degrees. A USDA article has issued a health warning about some tea bags that are sealed with a polyfluoroalkyl compound, PFAS, such as the chemical in Teflon cooking pans. It is evident in some teabags with thin edges that are not crimped, that the glue holding the paper together may contain this substance. I do not think our teabags have this gluing compound as the paper will work on either side. But the filter paper rolls are made in China, which do not have the health and environmental concerns that we do.

Tea Tours

I began doing bus tours for the Fairhope Museum in the 1980's. I had a large retirement village that would pay me $100 to "step on" their bus any time I wanted and talk about anything I wished. That money went to the Friends of the Fairhope Museum. After I moved to Lyter Ln it was not long before I regularly had them coming to the tea farm for a tea tour.

It was not long during this time when several Midwestern bus tour companies started stopping at the tea farm on their way to Orlando. There would be 55 people on the bus and I would charge them $5 a person for a porcelain cup and a chair. I would sell them some black tea and would do one or two of these tours a week.

When I retired from the museum in 2017, I started doing tours almost every day. Soon, bus tours and large groups got to be too much work. We did not like them using our restrooms and large groups notoriously do not buy much tea. So, we began turning down the retirement homes, garden clubs, master gardeners, school groups and bus tour companies. It took me over a year to stop the bus tour companies.

I now do small tours every day with just two or three people. I will take up to six but only three guests will fit on my golf cart. I will only do one a day at 1:00 pm and will get many calls a week from tourists who are "here today and tomorrow."

I do not really use a canned script for my presentations. I usually start out telling them where tea comes from and where I found my plants. After that, because I have been doing tea for 46 years, I can make it up as I go along. That does make it fresh and candid, but what I tell tourists varies greatly and evolves rapidly so that a guest who comes back again will hear a different program.

So here is a version of what I might say. Be patient as some of this material has already been written in previous chapters.

When booking the tour reservations, I ask for a first name (those help me keep them straight), how many folks might I expect (for amount of tea and number of cups on the tray) and a phone number in case of rain.

Before the guests arrive, I will start an hour out, putting on a fresh pot of tea. I have to be ready as they often arrive half an hour early because they have no idea where they are going. They will say "I'm sorry we are early," but I expect that.

"Hi, I am Donnie Barrett, welcome to the tea farm!" While standing by their car, "I expect you to be here an hour or hour and a half – depending on how many questions you ask – but if you get too hot or too cold, it will be shorter." I will go on to tell them to ignore the dogs and beware of garden spiders, or the many insects trying to get a drop of their blood. Also, I will tell them where the restroom is.

The Grandfather of American Tea: My Tea Story

"I'm about to serve you a cup of hot tea (I never serve iced tea), would you like a bit of sugar in your tea. I go on and bring them on to the covered car port, invite them to sit and say, "you can walk around, take pictures, make yourself comfortable and I will get us a cup of tea."

I will serve them the cups of tea, tell them what tea I have made and begin telling them about tea and where it originated and where I got my tea plants. I will wave around a large branch of tea and say "people are surprised when they find out all tea comes from this plant, *Camellia sinensis*. This plant comes from where Tibet and Nepal come together down to the top of Laos and Miramar. It grows as an understory plant and lives to be very old in a plant community just as Mountain Laurel grows.

The Chinese discovered them some 5,000 years ago, and brought them down into a semi-tropical climate where they grow into large bushes. I would then say, "In a few minutes you will see thousands of plants growing into big bushes."

I then tell them the story of how I recovered the three plants I started out with after recovering them from a burn pile. Because I had training at a commercial nursery at Auburn University I began reproducing the plants in large numbers. I had no idea what I was going to do with hundreds of tea plants or that there were very few people in the USA growing tea.

When I had produced an established patch of tea plants I quickly found out;

1. All tea comes from the same plant.

2. You don't just pick and dry the leaves.
3. Tea craft is a secretive process.

This is when I decided to go to China in 1984. I visited farms and tea factories all over China. They freely answered my questions and I actually stole technical secrets from the Chinese.

*When I took this picture these ladies had
no idea what type of tea they were picking.*

This is when I met my soon to be lifelong friend Shen Hua. He wanted to come to United States for an "American education." It was an adventure having him here for a year.

Next, I ask my tour guests "do you ever drink green tea?" Most do not and don't like it. Some say they drink green tea sometimes because that's the answer I want to hear. I will get out a box of green and black and say "green tea is green, Black tea is black." Calling tea green and black is brand new terminology our grandparents did not hear. I then give them a biochemistry class on the difference and how they are produced.

I tell them the story of how Mr. Lipton bought Sri Lanka in 1906 and started making black tea. He called it "tea" and started giving out samples in muslin bags. I tell them how the Boston Tea party was actually tea smugglers protecting their profits and they threw "Bohea" in the harbor.

Then we load up on the golf cart and I take them out into the tea field. I first stop in the "shade garden" area and tell them this is tea's natural environment under a canopy of large trees, growing as an understory plant. I tell them about the difference in the cellular structure of a leaf grown in the shade and one grown in the sun.

We will then go out into the field and see how we grow some of the rows in a round top, or "cupcake" shape and others in a flat or "pancake" shape. I will tear off a sample from each row and say "we make green tea from this in one hand, and black from this in the other."

We swing by the *Camellia assamica* bush and I tell them where it originates on the western side of the Himalayan mountains in the Assam states of India. "You'll see this in two places, as white tea and as a blend in breakfast teas.

We stop by the dry pen, and I will show them how we dry our tea in the sun. I will often time a batch of oxidizing black tea to be ready at this point, and spread it out from the little tub onto the old bed sheet. I ask them "how do you describe this smell" and am amused at their answers.

We end the tour with a look at my gift shop. A small tray holds bags of cup size teabags, green or black and half pound bags of loose leaf tea, green or black which they can purchase. "At this point you will owe me $20 per person which you will put in the gift shop box," ends the tour.

"Gift shop" has a sample teabag, some rack cards, two five-dollar bills for change and a cottage industry statement required at the point of taking money.

Tea Professionals

Being in the tea business for forty years has seen us attract all sorts of people to the tea farm. Folks with a high interest in tea will come long distances just to find you. Some are very nice and polite, and some come only to convince me on how much they know about tea. A couple of these guests have pointed out that I couldn't possibly know how to grow and make tea, not REAL tea, anyway.

The number of tea professionals who have come to visit our farm is vast. I have had professional tea blenders tell me how they conduct their craft. I have had tea buyers who fly the world over buying tea for their companies and their knowledge is more than impressive. I have had producers who work in the labs analyzing the flavonols to produce the process food teas called "cut-tear-curl" tell me how they do it. We have seen many distributors, shop owners and purchasing agents come to our farm to share their experiences. And the most amusing are the people who produce formal teas, "high teas" and Mother-daughter tea parties to come and show us the correct posture for curving the pinkie while holding a cup of tea.

When I first started making and selling tea I heard about the existence of "certified tea tasters." These seem to be of a British origin with students obtaining this certification from British schools. They were most knowledgeable about the subject. They could tell you the names of the different styles, companies and origin of hundreds of teas. They had words to describe the "nose," aromas, mouth feel, and the many subtle flavors such as "smokey melon, or buttered biscuit." Their understanding of tea was impressive and their vast knowledge of the subject often put me in the "Alabama white trash" category. Such as the French cuisine TV program they filmed here in 2003. The lady turned to me and asked "you do use bottled water, don't you?" I said "no ma'am, I just pump it out of the ground," pointing toward my pump house. For the rest of the program, all in French, she referred to me as "The Dukes of Hazard."

It seems now there is a new crop of these certified tea tasters coming from renowned academy's or from most every Ivy League US college. There is also an International Tea Masters Association which offers degrees in Certified Tea Master, Certified Tea Specialist, Certified Tea Connoisseurs and Certified Tea Sommelier. These certificates can cost thousands of dollars or you can buy the fifty-dollar book and take a quick course on line.

I have enjoyed meeting these tea professionals and have learned lots from them. The world of tea is eternally vast and learning more about tea never stops.

Shen Hua

When I first went to China in 1984, the country had just been opened to tourism by President Nixon's visit several years before. It was a Maoist country where adults wore only white with dark pants and skirts, and only children wore bright colors. Citizens did not speak loudly or complain out loud in public, they called it "loosing face." All this would quickly change with the new institution of capitalism. I thought it sad to see entire blocks of traditional Chinese structures being crushed for new modern buildings.

When I visited the "City of Gardens," Shuzou, China, I met a cook coming out of the hotel kitchen and he said "you have on pallamaine shoes." I had just bought some green tennis shoes and was intrigued that he had a name for them, so I asked him to repeat that name so I could write it down. "P. L. A. man shoes," he said. I laughed as it was People's Liberation Army shoes and we then and there became lifelong friends.

Shen Hua and I corresponded for several years, sending pictures, letters and gifts. He asked me once for a special, treasured gift that was unobtainable in China, a Walton Cake Decorator set. In 1994 he asked me to sponsor him to come to the US for an "American Education." So I signed up to take care of him financially, and pay for any needed transportation or health care. I enrolled him to take culinary classes at Bishop St. Jr. College in Mobile, and with a green card passport, he came to Fairhope, Alabama.

When he first arrived at the Mobile airport, my brother Ronnie picked him up and brought him to a grand reception at one of the homes on Government St. I had not even talked to him as he entered the room of finely dressed guests who cheered. He was led up to the hors d'oeuvres table where he tried one. He chewed it up then spit it out on the linen table cloth. That brought gasps, shrieks and cleared the room!

This was a very common practice and accepted custom in China. If you got fish bones in your mouth, you spit it on the table. I remember in China being seated at a nice restaurant with little chew wads from a previous customer around my clean dinner plate. So with this initial culture shock we began our year long venture with Shen Hua.

It was nice having him living at our house for a year. I would come home from work and a fire was burning in the fireplace, the mail was fanned out on the table, your clothes were washed, folded and put up and supper was on the stove. He could disassemble a fish or chicken and you could not recognize any of it. He would have a pile of fish scales, a pile of bones, a head and eyeballs on the side next to a clean pile of fish meat.

During the year, at the time he was attending Bishop State Jr. College for several months, he lived with my brother in Mobile. Ronnie gave him a nice room and had him work in the stage set business during Mardi Gras (a wonderful American experience) in exchange for his free room and board.

When he lived with us, he would divide the food into three parts and we would sit down to eat. One night I said this was not our Southern style. I said the cook would leave some on the stove, and after folks had eaten the first plate of food the cook would say, and I grabbed his chin and instructed, in Southern English, "Ya'll want some more?" When he went back to China and said that line of Southern English in a job interview at a new Pillsbury Shuzou plant, they fell off the stool laughing and he got the job which launched his baking career. He now is a quality control manager in the largest baking industry company in the world. He was always so appreciative because we gave him the boost to achieve this position.

With his name Shen Hua, we erroneously called him Shen, in the Anglican tradition, whereas his first name is Hua. Several years ago, because he does so much international travel and business he changed his name to "David Shen" and his wife's name to "Janet." We later had a chow dog we named Shen and David was highly insulted.

We have surprisingly stayed in touch and remained friends, now for 40 years. We speak warmly on FaceTime of our long international friendship although we do not always agree on much. Nothing is said about politics or world situations. He always thought leisure time was wasted time and he thought alcoholic beverages were detrimental to one's life and career. He has lightened up a little bit in his older age. He likes the way I use the English language and says our conversations are a good English lesson for him.

He has visited us several times over the years when he flies around the world. I always think his wristwatch cost as much as our house and he gives us advice on what he thinks we need to do. He was to visit in 2020 but said he could not fly because of some disease. I said then come later in a couple months and he said, "No, you do not understand, this is serious!"

In 2007, I returned to the closed and abandoned hotel where Shen Hua and I first met to pose for this photo. I was sitting on this step when he noticed my new P.L.A. man shoes, starting a lifelong friendship.

His descriptions of their Covid experience were much different than ours.

David's daughter, now "Tina," came to the US for an "American education" and was married and is now (Spring 2025) about to have a baby at her new home in New Jersey. I'm sure David and Janet will come to visit Alabama again on one of his trips to see their new grandchild.

My Life Story

Remember the Gladys Knight song which said "if someone should ever write my life's story, for whatever reason there might be…" Well, there is no reason to write my life's story and no one to write it so it looks like I am going to have to do it myself.

My parents, James (Bill) Edward Barrett, Jr. and Catherine Glasgow, met in Verbena Alabama in 1946.

This was my father's hometown where my mother found a school teaching job. They both had just gotten out of the army after serving during WWII. They married and moved to Auburn so my father could complete college and where my mother delivered my older brother Jimmy, James Edward Barrett III.

Bill Barrett was offered a job, after graduating, by the University to work in their experimental farm system and was offered a job in Cullman or Fairhope, Alabama. He chose Fairhope because he thought the name sounded nicer. They moved to Fairhope in 1948 with the position of Assistant Superintendent of the Gulf Coast Substation. They rented a house on Summit St. for a year while a new house for them was being built.

My twin brother, Ronald Glasgow Barrett, and me, Donald Hatchett Barrett, were born at Dr. Jordan's Bayside Clinic at Battles Wharf on August 16, 1949. It was the same hospital that my wife Lottie Jones was born in 1946.

Ronnie on left, Donnie on right. Aren't we cute?

My brother and I were fortunate to attend kindergarten at the Organic School. Then we attended Fairhope Schools and graduated in 1968. We attended Yancey Jr. College for two years then Ronnie went to the University of Alabama, my mother's school, and I went to Auburn, my father's school.

While we were growing up on the Experiment Station, a friend in my kindergarten class was Bill Simmons who became a longtime friend. His mother was Flo Simmons who drove us boys all over the Gulf Coast to see local attractions. Her hobby was walking the beaches, picking up Indian pottery and shells as us kids ran wild up and down the beach. One day I looked in her hand and she had a bullet, a button and an arrowhead. That moment changed my life and I became a beach comber and started learning lots about found artifacts.

Flo Simmons and I set up the first Fairhope Museum from her collection in 1995. She was difficult at times during the next several years not wanting the attention I was bringing to the museum. "We are a small country museum," she would say when I was planning a feature show.

She died in 2005 and would have not been able to stand the transition to the Fairhope Museum of History in 2007. When she died, I put in her hand a bullet, a button and an arrowhead. I always thought I would write a book of that title honoring Flo and how she changed my life and set me on the course of what I was to become. I'm sorry Flo, but I never did.

While I was still living at home I started a persimmon grafting project. Behind the dairy barn at the substation there was a wooded area with lots of mature wild persimmon trees. I started grafting Chinese persimmons up in the branches of the wild trees. Within two years I had an amazing stand of Chinese persimmon trees. Just as they were producing fruit a few years later, the substation cleared the land for some manure spreading experiment.

Also, during this time, my girlfriend bought a matchbox of marijuana which had a few seeds in it. I stuck a seed into one of my mother's house plants which sprouted. I took it out into the woods, and it grew into a twelve-foot bush. I had no fear of anyone finding it, but soon it became evident that it was not "worth the trouble" to grow pot plants.

My first year at Auburn I took pre-veterinarian classes, which I later changed majors because it was too hard (big mistake #1) and studied Wildlife Biology, which was more fun. My wife Sheila became pregnant during my senior year so we moved back to Fairhope and I later graduated from the University of South Alabama. My daughter, Angela Shelby, named for my wife's college friend and a local lake, was born in 1974.

While at Auburn I started working at a funeral home in nearby Opelika. I told them that I had experience assisting with autopsies in Fairhope which got me the night job of Autopsy Assistant at the Alabama Toxicology Unit in Auburn near the vet school. It was the only forensic lab in Alabama at the time and we would do crime involved autopsies from all over the state, usually several a night. I got real good at it. I worked there for two years and benefited greatly from the experiences, medical training and cases of human drama.

Also, while at Auburn I started doing yoga. A young man on a wildlife field trip told me about a "yoga house" while we were out standing on a beaver dam. The house was only a few doors down from where I lived so one evening I walked in and sat down and listened to the several teachers. I have practiced yoga and tai chi my entire adult life, even teaching it for a few years at the Oriental Institute in Barnwell.

When we arrived back in Fairhope I found a job at Poser Business Forms, where my mother-in-law was employed. I thought I would work there three months, and it turned out to be 31 years. It was quite a culture shock for me to be immersed in a group of completely uneducated people; they were involved in fights, wife beatings, low class people who knew little about anything. There were a few good people, such as my second wife Lottie who worked there. We dated for 14 years and were married in 1991. This job allowed me to work at an easy, mindless job, get well paid and have lots of time off to pursue many avocations.

After Lottie and I went to the King Tut exhibit in New Orleans in 1977, I found I could draw and paint Egyptian hieroglyphic characters. We then scheduled a trip to Egypt in 1978, where I fell in love with Egyptian art, sculpture and hieroglyphic writing. I even achieved the ability to read common phrases.

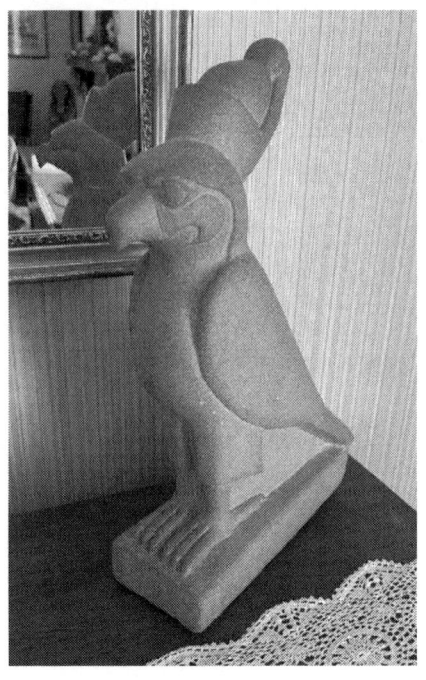

I made sculpture in limestone with a hammer and chisel.

For the next three years I painted Egyptian wall paintings and for years after sculpted limestone copies of famous Egyptian sculptures. To this day I do a sculpture in marble every year of what happened that previous year.

The people of Fairhope accepted me covering an entire building in Egyptian art because Fairhope is such an artistic town.

From 1984 to 1992, I was appointed to the Baldwin County Archaeological Advisory Review Board. Our job was to research archaeological and historical sites that were under development or endangered by encroachment. We were then supposed to report back to the Baldwin County Commission on sites conditions and threats. They sent us to classes in Montgomery such as archaeological law and historic preservation techniques.

I became a historic preservation shark! I was empowered by being educated on the subject and living in a very historic county being consumed with development. Rarely did we report back to the commissioners. We would attend their meetings, and they cared little of our concerns and continued to support all development. That is what "commission" means.

A couple of us on the committee took it upon ourselves to use the tools of the Alabama Historic Commission in Montgomery to inform developers when they were breaking the law. We would report such to the AHC and they would send the developers threatening letters. We saved several Indian mound systems and historical battlefield sites. During this time (1995) I single handedly stopped the development on the 1780 original Spanish Fort. This is one of my proudest life-time accomplishments.

In 1984 I participated in a movie "First Frontier" in Talladega, AL. It was an 1812 movie about Andrew Jackson winning the 1813 Creek War. I saw them tie red rags on college students to make them Creek Indians and thought I could do that at Fort Mims. I went to the Ft Mims Restoration Association and proposed a reenactment. They were cool to the idea but let me arrange one. We used Civil War reenactors with percussion cap guns and did the program on the original fort site which had become a garbage dump. Now, forty years later the park is beautifully furnished, the fort rebuilt, and the reenactment has grown to be an Alabama tourism premiere event on the original site. I am given the credit for starting it all.

I had worked at Poser Business Forms for 31 years. It was an easy job that they paid you nicely just to be there every day. They gave me lots of vacation time so I could travel the southeastern United States going to Civil War reenactment events sometimes every weekend. That job ended in 2004.

Lottie and I went to a Civil War encampment at Fort Morgan about 1982 or 3. I was instantly hooked. I saw the potential in these guys to help produce a living history event. I joined the reenactment group and learned the ropes. In one year, I took over its leadership and became the captain of the 21st Alabama Infantry.

I immediately started making up and organizing events where they had never had any type history show before. This included Oven Bluff, Jackson Bluff, Blakeley, Citronelle, Silverhill and that 1813 event at Fort Mims. We did parades, CW dances, stood around as "eye candy" at old plantation homes like Oakleigh and did funerals and lots of CW weddings. We made Fort Morgan many thousands of $$'s with huge reenactments every year and when Fort Gaines was in financial trouble, put on an event every month for almost a year. Some of these were impromptu C W plays. I became a leader of the whole "hobby" and was known as "the clearing house for reenactor involvement."

This all came tumbling down for several reasons. The Confederate Flag issue split the reenactment community in half with the educated historians on one side and the yahoo rebel flag wavers on the other. Also, the groups or units had no real glue to hold them together. Members would drift from one unit to another with no loyalty to any. That became difficult to manage.

I remained the struggling leader for a year in 1999 because my brother asked me stay the leader so we could have a Civil War themed birthday party in August. So I did. We had a grand two-day event. But, on the first night at the Nix Center in Fairhope I was organizing the several stage shows and music groups. Just moments before the last act was done and the country music band began (I had just queued up) it happened. While standing in the middle of the ballroom floor juggling food, people, bands and stage acts, I suddenly left the dance floor to the utter astonishment of all our assembled family, friends, coworkers, classmates and the entire reenactment community.

This is Lottie's family at the Civil War party in 1999. Lottie's father was not there as he thought it all was nonsense.

So, at that point I quit the Civil War business. My last event in a Confederate uniform was the Dec 31, 1999 sunset program that I made up standing on top of Fort Morgan, in front of thousands of people, talking about "time" as a perfect sunset closed out the century and my Civil War career.

After my Civil War period ended, in 2004, I started teaching at the Organic School. Big mistake! I told them up front that I did not like children and believe me, they did not like me. I did so much for the kids there, hands on projects every day which I designed and produced the materials during the summer for the next year. The school was full of troublemakers expelled from public school and they focused on me. They lied to their parents on how bad I was and destroyed my projects and materials and just laughed in contempt. I was even the school board president for a year and all we did was disagree. My management style did not work there and I could not save them from their own self destruction. I was so glad to leave when I quit in 2007 to become the Fairhope Museum director.

I was the volunteer director of the Fairhope Historical Museum for twelve years when Mayor Kant offered me the City job. This started a new department in the City network which no one could relate to. Department heads were then paid around $90,000 a year but they only paid me half of that – which was corrected a couple years later. It was better than the $18,000 that I was paid at the Organic school.

I worked for the City for eleven years. It had its good times and its bad times. The facility I built was way under budget and amazingly good and very well received by townspeople, City administrators, museum professionals and thousands of visitors. The weekly programs I put together were done so well and we put on a festival of some type every month. I decorated the museum exterior with huge decorations for every season and holiday. The Friends organization was overly active and paid for most everything I could dream up. The City's purchasing dept constantly gave me fits. I signed my own purchase orders, bought from whom I wished (not the approved vendors), quit even writing purchase orders at all. Then the City, impressed with my frugal nature and stunning success, gave me "executive privilege" to purchase as I wished. The Museum became Fairhope's showplace and I became "Mr. Fairhope." I was constantly bathed in praise, people would call out my name from a crowd, which felt really good!

When I turned 69 it became time to retire. I had to surrender my place of high esteem and be put out to pasture. I was a bit startled how quickly people turned their back on me, how the respect stopped and how many of my past museum Friends became so jealous and ugly. The streak of museum directors after I left told me they did not need me to come to the facility, or I needed to make an appointment first, and they didn't need my help with anything. A big mistake was leaving all my library, files and research materials there thinking I would certainly be welcomed to use my own materials for research projects, but I did not feel comfortable doing so.

When I was working, I joked many times saying "the tea farm will be my retirement!" Well, that became the fact. I did not have to plant any more bushes, had all I could manage, but started doing tea farm tours on a daily basis. Lottie and I have worked every day on tea, doing tours and making the sales units for customers.

When I first started organizing the Fairhope Historical Museum in 1995, at hand I had all this original material that had been donated to the ladies who were trying to start a museum. I found early Couriers, letters from Bellangee, Gaston and Marie Howland and lots of early documents. These mentioned "cooperatives" along with populist publications and related letters. I put together, after some time of thought and reading, that Fairhope was meant to be a Populist, cooperative colony and single tax theories came along later.

When I graduated from University of South Alabama in 2000, my senior thesis was "The Fairhope Industrial Association 0f 1894: Putting Populist Ideas to Work." Then later I lead an effort to get the Single Tax Corp. to change their information publications which stated "Fairhope was founded by the De Moines Single Tax Club who came to Fairhope to prove the principles of Henry George." While I was the Fairhope Museum of History director, I gave out stacks of one-page flyers "Henry George or Populism?" stating Fairhope was planned and executed by the populist members of the Fairhope Industrial Association. This also stated nothing from that time mentioned single tax theories until years later when they were trying to win some financial support.

In 2022 I did a tongue-in-cheek speech and paper titled "The Fair Hope is Over," stating the hopes and dreams of the original founders who had a fair hope of succeeding is over because of the commercial City that stands here now. People could not follow my line of thinking. When I bought a paver for the 1925 High School I put on the bottom that "The fair hope is over" and they put on the paver, "Fairhope Is Over," not at all what I was trying to say.

In 2024 my campaign for truth is again put to the test. Developers are about to tear down the old Fairhope Hardware building on Fairhope Ave. While removing layers of roofing, façade material and paint, they have uncovered the original 1922 sign of the "The People's Cooperative Store." I wrote a short article suggesting folks just look up and appreciate the eight-foot-long word "cooperative." The article stated that the founders of Fairhope said Fairhope was to be based on cooperative distribution, and they had a fair hope of establishing a model community. This 1922 building was their last cooperative effort. When that sign, the last speck of their founding history is gone, then truly their fair hope is over. I received much positive support and feedback from preservationist minded Fairhopeans.

Well, I'll end this here, but I do hope I have many more fruitful years yet to go.

This is 2025 and the original tea bushes I recovered from the Lipton expirement 46 years ago.

Conclusion

The Happy Plucker!

In conclusion...I have written a memoir or review of how I have now found myself in the tea making business. It is not meant to be how tea is grown, produced and brewed but how I have learned to do it. In my conversational tone, in my own rather eighth grade voice I have described my mistakes, accomplishments, my suppositions, experiments and outright failures. But, we have arrived over a bumpy road to making a very good cup of brewed tea.

We have never thought of ourselves being in competition with anyone. I really seem to do things to please myself and make the teas the way I like them with little consideration to any "market." As I have not depended on income from tea sales, this has always worked well for me.

Upon reading and talking with the other producers, I've known, my processing barely resembles how others make their teas. We have very little in common except we start out with a green leaf. It is almost amusing to see our conflicting ideas on oxidation, caffeine, when and how to pluck, and how the different flavors develop. Most of these others will throw this book across the room!

In 1979, I was given the three original Lipton plants. I was not looking for them or thinking about growing tea, and I actually remember little about the day I received three tea bushes in the back of my truck. Within a few years I had lots of seedlings sprouting under my bushes which I did not need so I would pull them up and give them away. I would jokingly give people three seedlings and say, "I started my farm with just three plants." I still give tea tour visitors 3-10 plants for free.

Soon I had people coming to just get the plants, hundreds at a time and then thousands at a time. So, I started charging a dollar each just to compensate my pulling them up and packing in wet paper and double plastic bags, but at this price, this is still giving them away. When I would get a young couple who wanted to start a tea-producing garden and scrapped up $500 for 500 plants, I would give them 1000 plants and charge them 0-$200. I have made many tea growing friends doing that.

I still charge only $1 for a seedling. Another tea growing friend of mine calls my tea plants "Fairhope Select" and charges $34.95 for a three-year-old plant. I seem to get lots of tea plant business selling 3-5000 plants a year.

I give my tea and fresh eggs to a friend, Jon, when I buy his honey. Not long ago he said, "you are so generous." I replied, "it's the way to be." Another lifelong friend, Jahala, really gave me a compliment saying, "the most successful people give it all away!"

The presumptuous title of the book, "Grandfather of American Tea," which I am going to regret using some day, is bound to bring heckling and scorn from other tea growers, but I have used it simply because of my early entry into growing and making tea. I certainly have no influence or bearing on any of the many other American tea growers, who have never heard of Fairhope Tea, and certainly do not consider myself a leading tea industry pioneer. I just thought the name had a nice ring to it. I feel a sensation of letting others read my personal diary with this book. It says what I know about tea, but also reveals what I don't know about tea but here it is presented for your approval.

One chapter that is not included here was titled "Great Stories." I was moving it from one place to another on the computer and lost it. I never could recover it which was too bad because I could not rewrite it. For two years I recorded the many wild and crazy things that visitors would say. From a Proud-Boy type saying he was "going to take me out if I was trying to poison his family," to the gentleman who asked "do you buy instant tea to put in your tea so it will taste like tea?" and the man from India who came to visit, would not try my tea, would not pay for the tour then went on to assure me I was not making REAL tea, I've had them all.

Another chapter that is missing is a log of where I have sold plants, and which farms got their start from me. I started giving away bundles of bare rooted seedlings in the mid 1990's. Since then, I have sold many thousands of plants, for less than a dollar, to farms from California, North Idaho to Virginia and the Carolinians down to the tip of Florida. I did not want to have to contact all these people for permission to use their names.

Well, that's about it! Donnie out.

Acknowledgments

The "Grandfather of American Tea" sounds quite presumptuous, doesn't it? This name came from a long-time tea friend, James Oorock, Ph. D., of Single Origin Teas, Inc. (he sells my tea by the gram!), who first published it on his website many years ago. The title is not at all a completely accurate claim, but can be applied to the fact that I started growing tea before most all of the other American tea producers. And several southeastern tea growers received their first plants from Fairhope, and the information on how to make them grow, then three years later how to make the finished product.

I would like to thank my other tea making friends who have helped me down through the years.

Dr. Mike Loeb, Ph.D., who lives in Tallahassee, FL, has been a friend for many years. He is highly educated and an experienced tea-taster. He thinks like I do and freely shares his experience and knowledge. His advice has been extremely valuable and we have become good friends discussing problems, thoughts, techniques, and experiments. We don't always agree but that is what you need. Having someone so knowledgeable, who is available on a daily basis as needed to run a problem by, has been a pillar of strength for me to lean on in my tea journey. A true friend, thank you, Mike!

Col. Jason McDonald, of The Great Mississippi Tea Company in Brookhaven, MS, is always someone you can call on. He is most knowledgeable on growing tea techniques and the use and sources of tea making equipment. His farm is vast and he has experienced it all! He will answer any question posed to him, and will share information freely on his Facebook site "Let's Grow Tea." He and his partner Tim teach tea making classes online. I can always depend on Jason for the right information at the right time.

Steve Lorch, RN/BSN, owner and founder of the Table Rock Tea Company, is the man to go to when you need any equipment to produce tea. He is also an excellent tech support person who will help you solve any problems with the machinery. His book, "How to Grow and Make Tea in the United States 2nd edition" is an excellent read.

Catherine King of Fairhope has helped me for many years, especially while I was director of the Museum. She would fact check what I was saying/doing/producing and always has kept me on tract.

Jeanine Normand has been a great fan of our Fairhope Tea from the beginning days. She has promoted our tea in many ways. Her friendship was sealed when we both were in the fifth grade and dancing on her mother's front porch. Our song "Soldier Boy" was playing and I nervously kissed her on the cheek!

Most of all I want to thank my wife Lottie. She was skeptical of my tea growing for years but later became a team player. Our operation is a two-man, full time job and

I would have never sold the first ounce of tea without her constant help and support.

I invite you to visit our website:

https://fairhopeteaplantation.com/

Made in the USA
Las Vegas, NV
19 July 2025

106117e4-4ad1-4966-8603-8c6fc8a98762R01